*Certainty and Uncertainty in
Biochemical Techniques*

Certainty and Uncertainty in Biochemical Techniques

HAROLD HILLMAN
M.B., B.Sc., Ph.D.

QH324
H54
1972

SURREY UNIVERSITY PRESS

HENLEY-ON-THAMES, OXON.

First published 1972

© Harold Hillman, 1972

All rights reserved. No part of this publication may be reproduced, stored in a retrieval system, or transmitted in any form, or by any means, electronic, mechanical, photocopying, recording or otherwise, without the prior permission of the Copyright owner.

ISBN 0 90 3384 00 0

Typeset in Great Britain by
Santype Limited (Coldtype Division)
Salisbury, Wiltshire, England
and printed by
Alden & Mowbray Limited
at the Alden Press,
Oxford, England

We have seen that a very considerable quantity of heat may be excited in the friction of two metallic surfaces, and given off in a constant stream or flux, *in all directions*, without interruption or intermission, and without any signs of diminution or exhaustion.

Rumford, B. (1798), *Phil. Trans. Roy. Soc.*, 88, 80

In the development of separation methods, a critical consideration is that the products, often labile in character, shall not be damaged by the processes which lead to their isolation as entities, a requirement frequently necessitating unremitting vigilance. In this context, physical methods present many advantages for isolating the more complex components of the cell.

Kekwick, R. A. (1966), *Brit. Med. Bull.*, 22, 103

easdem coram populo in eminenti loco comburi faciant ut hujusmodi punitio metum incutiat mentibus aliorum ne hujusmodi nephande doctrine et opiniones heretice et erronee vel ipsarum auctores et fautores in dictis regno et dominiis contra fidem.... quod absit sustententur seu quomodolibet tolerentur....

Anno secundo Henrici IV (1400), 16, 441

Dedicated to Elizabeth
and all other good people
urbi et orbe

Contents

	Preface	ix
1	Subcellular fractionation	1
2	Histochemistry	41
3	Electronmicroscopy	54
4	Radioactive measurements	67
5	Electrophoresis	79
6	Chromatography	89
7	General characteristics of all techniques examined	96
8	The individual techniques	98
9	The definition of biochemistry	103
10	Practical conclusions	108
11	Proposed strategy for biochemistry	110
12	Summary and conclusions	114
	Appendix 1: Misleading synonyms	115
	Appendix 2: Radial pressure distribution in a rotating liquid	121
	Appendix 3: Temperature change for a pure substrate during a reversible adiabatic pressure change	123
	Index	124

Preface

This study arose out of some thoughts about the relevance of the experiments on friction carried out by Count Rumford in the 1790s to current techniques of subcellular fractionation. I gradually came to the conclusion that an analytical look was needed at the whole procedure, to see what unspoken assumptions were implied by its employment; it then became necessary to examine their validity.

Having established the general approach, I took five other popular techniques used in animal biochemistry, and subjected them to a similar analysis. This consisted of (1) describing all the steps in a procedure; (2) examining the agents in each step which might influence the final answer; (3) identifying the assumptions necessarily implied in the use of the procedure; (4) discussing their validity; and (5) suggesting control experiments to analyse quantitatively how each step affects the conclusions of experiments. At the same time it became necessary to classify the meanings of the term 'biochemistry', (see p. 103).

Enzymes from animal sources were used as the model compounds for the following reasons: they are characteristic of living tissues; they are proteins; they are relatively unstable; they are widely studied by these techniques; and there is plenty of experimental data in the literature about them.

The present form of analysis could, of course, be applied to any biological technique, and it is hoped that it will be.

Undoubtedly, a great deal of useful biochemical knowledge has been accumulated by careful experiments carried out with full appreciation of the limitations of the techniques used. However, with the development of increasingly elaborate biochemical procedures, there is the inevitable risk that the interpretation of experiments may reflect the technique more than the fundamental properties under investigation. The purpose of this book is to analyse critically the factors upon which interpretations are based.

I will be most grateful for serious criticism, or supporting evidence, for the ideas expressed. I would be particularly obliged to anyone who would question that any of the assumptions listed are not implied by the use of the techniques. I undertake to reply to all serious criticisms.

The original impetus for these studies came from some extremely careful and painstaking experiments carried out by Mrs Anita Palm of the University of Gothenburg, Sweden, to whom I am most grateful. I have much pleasure in thanking Mr Donald Fargie of the Department of Mechanical Engineering, Surrey University, for deriving the equations in the Appendices, and Dr Zdenek Ernst, of the Department of Chemistry for many very valuable discussions and much sound advice. Dr Leif Hertz, Miss Babette Stern, Professor Robert Thomas and Dr Maurice Moss, kindly read the manuscript and made many useful suggestions. The responsibility for the views is, however, entirely my own.

Harold Hillman, Unity Laboratory,
Department of Biological
Sciences, University of Surrey.

I
Subcellular Fractionation

Subcellular fractionation consists of the following steps, each of which will be examined separately:

1. Killing the animal.
2. The dead animal cools.
3. The tissue is diluted.
4. The tissue is homogenized.
5. The preparation is added to a sucrose gradient.
6. The homogenate is centrifuged.
7. The substrate mixture is added.
8. The products of the enzymic reaction are extracted.
9. The products of the reaction are measured.

1. Killing the animal

Killing is usually carried out by severance of the neck by guillotine or traction (Vintner, 1966), or by rendering the animal unconscious by a blow on the neck and then exsanguinating it by cutting the carotid arteries (McIlwain & Rodnight, 1964, page 1). Probably when the animal is restrained before being killed, it has a stress reaction; this may come from sudden apprehension, or it may arise later from pain receptors in the neck during traction, just before the animal loses consciousness. Certainly the stress causes profound somatic and visceral change (for reviews, see Selye, 1951-56). The degree of biochemical change will presumably depend upon the time between the neck severance and the moment at which the blood supply to the adrenal cortex and other organs involved becomes insufficient for the stress reaction to continue.

The length of this interval would depend on the previous handling of the animal. For example, an animal not used to being handled or restrained by an experimenter might be more rapidly frightened and have a much earlier stress reaction. The time would also depend upon the blood flow not only to the adrenal gland, but to the whole autonomic nervous system involved in stress. The blood flow varies from second to second and depends upon the species of the animal, the physiological state of its heart, the nutrition of the animal, the relative sizes of the different organs, the redistribution of blood within a failing circulation, the size of animal, etc. Thus the degree of the stress reaction, and the biochemical consequences of it, are extremely difficult to measure or assess in a meaningful way. It might be feasible to define in respect of a particular strain of a particular species over a small age range with a known regime. From a practical point of view, however, such a procedure would probably be too complex and time consuming to be useful.

In animal experiments it is difficult to distinguish between the effects of stress and those of muscular activity, since the former usually induces the latter. Therefore, when discussing the problem it is easier to look at the effects of exercise, which can be measured quantitatively. It would probably be possible to dissociate them by studying the biochemistry of passive exercise in anaesthetized animals, when it may be presumed that stress is not occurring.

When unanaesthetized animals are killed, neck fracture is always followed by considerable muscular activity more pronounced in the hind limbs than the forelimbs (Feldman & Hillman, 1969). They pass through a short generalized spastic paralysis. Muscular exercise produces gross biochemical changes in the blood gases, its ionic concentrations – indeed in the overall body chemistry. A few of the important ones, with their degree of change are given in Table 1.

Changes quoted are those reported for mammals at a body temperature of 37°C. In the dying animal they would be compounded by the effects of dying itself. As far as the author is aware, most work on the biochemistry of dying has been published in the Soviet Union (see, for example, Gaevskaya, 1964; Negovsky, 1966). There is a monograph on the chemistry of death, (Evans, 1965) and notes on the pathology of the postmortem state (see, for example, Jellinek, 1947; Anderson, 1961). It is difficult to examine the blood biochemistry during

	Degree of exercise	Change	Reference
Arterial P_{O_2} (mm Hg)	Moderate	94 to 73	Lilienthal et al. (1946)
Alveolar P_{CO_2} (mm Hg)	Strenuous	39 to 45	Bannister et al. (1954)
Arterial pH (units)	Moderate	7.38 to 7.30	Hickman et al. (1951)
Blood lactate (mg%)	Strenuous	10 to 100	Bannister et al. (1954)
Plasma adrenalin (ug/L)	Moderate	0.07 to 0.22	Vendsalu (1960)
Plasma nor adrenalin (ug/L)	Moderate	0.42 to 0.78	Vendsalu (1960)

Table 1. Changes of blood or alveolar constituents resulting from exercise

dying because in this situation — as in hypothermia (Rogers & Hillman, 1970) — there is a redistribution of blood from the arteries in favour of the venous system, (Feldman, H. & Hillman, H. 1968, unpublished observations); sufficient blood can only be drawn from the larger veins, and there is little reason to assume that this blood is typical of capillary blood in equilibrium with the tissues, as is presumed to be the case *in vivo*. Arterial samples of sufficient quantity can be drawn only from large, costly animals. An analytical study of the biochemistry of dying would seem to be an urgent priority for research workers with sufficient resources, in view of the fact that tissues for study *in vitro* are taken from recently dead animals.

If one considers the changes of blood chemistry summarized in Table 1, it is immediately obvious that they would profoundly alter the course of many of the chemical reactions which one would like to follow in the subsequently isolated tissue. For example, the hyperventilation which occurs at the beginning of exercise, increases the oxygen and decreases the carbon dioxide tensions of the blood, and subsequently of the tissues; therefore the rate of oxidative phosphorylation – or any other reactions whose rates are limited by the amount of oxygen available – will be altered. Similarly a change in intracellular sodium ion concentration would alter the rates of the sodium activated ATPases (Skou, 1957; 1964). In the general case, there need be no concern if exercise alters the rate of a reaction before or during the death of the animal, as long as the rate returns to its previous rate during the incubation of the tissue. Nevertheless, we must first demonstrate this experimentally. For example, if the glucose level is depressed by exercise, its presence in 10 mm concentrations in the Krebs-Ringer medium will provide adequate substrate.

The corollary of this is that one should add catecholamines, steroids, amino-acids and hormones, not only to any final incubation medium of a tissue slice, but also to every stage of every homogenate, centrifugate, etc. This will preserve the right environment throughout, although it might make extractions and chemical estimations very difficult. Nevertheless, there would seem to be an incontrovertible case for adding chemically or hormonally active agents in physiological concentrations to all *final* media coming into contact with tissue. This may be a counsel of perfection.

The critical question is – how much are the particular changes occurring during the killing of the animal, or its dying, or in the preparation of the tissue, reversible *in vitro*? There is, of course, no general answer to this question. One usually adduces such criteria as linear respiration of tissue slices, like retina or brain (Warburg, 1930), or electrical activity in giant axons (Hodgkin, 1951), or a high level of phosphocreatine, as in cerebral slices, (Kratzing, 1953). It is pertinent to point out, however, that such criteria are but rarely used in respect of tissue homogenates, or 'sub-cellular fractions'. Objectively, experimental results should always be validated by several independent criteria of the viability of the tissues as compared with their state *in vivo*.

Animals may be anaesthetized before killing in order to avoid both stress and physical activity. If one assumes that the stress of the inhalation or injection of anaesthetic lasts for a few minutes, it might be worthwhile to wait, say, ten minutes, before cutting the carotid arteries and exsanguinating the animals. It would be preferable to wait until the heart rate and respiration of the animal had returned to its level *in vivo*, but this might be a very long time.

Here we are beset by the likelihood that the anaesthetic itself may alter the biochemical state of the animals. Anaesthetics generally lower the animal's temperature, albeit slowly. They seem to have the effect of raising the high-energy phosphates in the brain (for summary, see McIlwain & Rodnight, 1962, page 340). There are two interpretations of this finding. The simpler one is that tissue metabolism is decreased and, therefore, less of them are used up. The other explanation is that the anaesthetized animal is less stressed, and indulges in very little muscular activity when thrust into the liquid nitrogen, necessary for measuring the labile compounds; their level has thus not been lowered in the struggle (Stone, 1938; Dawson & Richter, 1948). There seems to the author to be no experiment which could decide between these explanations. If phosphocreatine and ATP are increased by anaesthetic, and there is no other practical way of avoiding the stress and muscular activity accompanying killing, then we have to share the current vicarious hope and assumption that the anaesthetic has no irreversible effect on the final measurement.

A very fundamental difficulty in studying tissues taken from dead animals is that the probable immediate cause of death — irrespective of the agency used, — is hypoxia, resulting from failing circulation. There is a large literature on the effects of hypoxia (see, for example, Walker & Turnbull, 1959; Dawes, 1968; Vogel, 1970). Lack of oxygen may cause irreversible changes extremely rapidly in normothermic animals. Minute vacuoles have been detected by electronmicroscopic examination of the brain within a few minutes of hypoxia (Brown & Brierley, 1968). One minute's delay in artificial respiration can cause the heart of a rat to stop in 12 minutes, compared with 83 minutes, if artificial respiration was started immediately (Feldman & Hillman, 1969).

2. The dead animal cools

As soon as a mammal is killed, it begins to cool down from a temperature of 37-39°C to the laboratory temperature of, say, 18-24°C. The rate of cooling depends upon the animal's initial temperature, the ambient temperature of the room, the size of the animal, its heat conductivity, the specific heat of its different tissues, and the heat generated by the autolytic and enzymic reactions continuing within it. The experimenter often accelerates the cooling by incising the animal, and then immersing the tissue in ice-cold trichloracetic acid, or to much lower temperatures with solid carbon dioxide or liquid nitrogen. The aim is to slow down the enzyme activities involved in tissue metabolism. Cooling to much below zero centigrade causes ice formation with disintegration of tissue; it cannot be used for studying metabolism in isolated organs, or tissue slices, but may be used for specialized single cells like bull sperm (for review, see Meryman, 1966). Where tissue integrity is not necessary, as in subcellular fractionation or enzyme preparations, more profound cooling may be undertaken.

The temperature coefficient of enzymes between 0°C and about 40°C is 2.5 to 3 per 10°C, so that a tissue of 0°C will metabolize about one thirtieth to a fortieth of its rate at 37°C. In neurosurgery, cardiac surgery, artificial insemination and food preservation, among many other examples, cooling has been shown to have a preservative effect on tissue. Its depression of the metabolism presumably permits the oxygen in the tissue to last longer. Yet, if different enzyme systems, transport mechanisms, membrane permeabilities, and non-enzymic reactions all have different temperature coefficients, the cooling would engender different relative new rates at a particular temperature. This would not matter when the biochemical preparations were subsequently rewarmed to 37°C for incubation, if they all returned to approximately the same rates as previously *in vivo*.

The question then is raised of how much irreversible change is induced not only when an isolated enzyme system is cooled, but when an enzyme in a multiphasic cycle is cooled. This would be relatively easy to examine and should be done. It is already well known that temperatures around and below zero may decrease enzyme activity and this has been called 'cold denaturation'. For example, 'mitochondrial' pyruvate decarboxylases from rat adipose tissue, chicken liver, and rabbit and rat mammary glands can increase their specific activity up to six-fold following freezing and thawing (Ballard & Hanson, 1967; Utter & Keech, 1963; Gul & Dils, 1969). The same enzyme from rat adipose tissue and kidney is much more stable at room temperature than around 0°C (Henning, Stumpf, Ohly & Seubert, 1966; Ballard & Hanson,

1967). The possible occurrence of this phenomenon at temperatures above and below zero used in routine biochemistry, in respect of single pure enzymes, enzyme preparations, tissue slices, etc., should be examined. Different systems would not necessarily give the same answer, but systematic examinations of all systems commonly used is surely desirable. Environmental, biochemical or mathematical corrections could then be made, if necessary.

The effectiveness of freezing in the preservation of such ephemeral properties as the capacity of bull sperm to fertilize, the tastes of fresh fruit, the infectivity of viruses, would all indicate that, in general, the cooling of tissues is unlikely to be significantly deleterious, but quantitative statements necessitate experimental validation of findings.

Anderson, W. A. D. (1961) in *Pathology*, 3rd edn. ed. by Anderson, W. A. D., St Louis, Mosby, page 90.
Ballard, F. J. & Hanson, R. W. (1967) *J. Lipid. Res.*, 8, 73.
Bannister, R. G., Cunningham, D. J. C. & Douglas, C. G. (1954) *J. Physiol.*, 125, 90.
Brown, A. W. & Brierley, J. B. (1968) *Brit. J. Exp. Path.*, 49, 87.
Dawes, G. S. (1968) *Foetal and Neonatal Physiology*, Chicago, Year Book Publishers.
Dawson, R. M. C., & Richter, D. (1950) *Amer. J. Physiol.*, 160, 203.
Evans, W. E. D. (1965) *The Chemistry of Death*, Springfield, Thomas.
Feldman, H., & Hillman, H. (1969) *Brit. J. Exp. Path.*, 50, 158.
Gaevskaya, M. S. (1964) *Biochemistry of the Brain During the Process of Dying and Resuscitation*, English trans. New York, Consultants Bureau.
Gul, B., & Dils, R. (1969) *Biochem. J.*, 111, 263.
Henning, H. V., Stumpf, B., Ohly, B., Feubert, W. (1966) *Biochem. Z.*, 344, 274.
Hickman, J. B., Pryor, W., Page, O. P. & Atwell, R. J. (1951) *J. Clin. Invest.*, 30, 503.
Hodgkin, A. L. (1951) *Biol. Revs.*, 26, 339.
Jellinek, S. (1947) *Dying, Apparent Death and Resuscitation*, London, Longmans.
Kratzing, C. C. (1953) *Biochem. J.*, 54, 312.
Lilienthal, J. L., Riley, R. L., Proemmel, D. D., & Franke, R. E. (1946) *Amer. J. Physiol.*, 147, 199.
McIllwain, H., & Rodnight, R. (1962) *Practical Neurochemistry*, London, Churchill.
Meryman, H. T. (1966) in *Cryobiology*, ed. by Meryman, H. T., London, Academic Press, page 1.
Negovsky, V. A. (1966) *Advances in Resuscitation*, Moscow, Soviet Academy of Sciences.
Rogers, P., & Hillman, H. (1970) *J. Appl. Physiol.*, 29, 58.
Selye, H. (1951-1956) *1st-6th Annual Report on Stress*, Montreal, Acta.
Skou, J. C. (1957) *Biochem. Biophys. Acta.*, 23, 394.
Skou, J. C. (1964) *Prog. Biophys. & Mol. Biol.*, 14, 131.
Stone, W. E. (1938) *Biochem. J.*, 32, 1908.
Utter, M. F., & Keech, D. B. (1963) *J. Biol. Chem.*, 238, 2603.
Vendsalu, A. (1960) *Acta. Physiol Scand.*, 49, Suppl. 173, 57.
Vintner, J. (1966) *Kind Killing*, London U.F.A.W.
Vogel, J. M. K. (1970) ed. *Hypoxia, High Altitude and the Heart*, Basil, Karger.
Walker, J., & Turnbull, A. C. (1957) eds. *Oxygen Supply to the Human Foetus*, Oxford, Blackwell.
Warburg, O. (1930) *Metabolism of Tumours*, trans. by F. Dickens, London, Constable.

3. Dilution of the tissue

Tissue is normally homogenized in an aqueous solution with or without a buffer. The effects of these additions are:

(i) dilution of all the soluble materials within the tissue;
(ii) chemical reaction of the diluent with the tissue;
(iii) swelling of the tissue;
(iv) loss by the tissue of soluble components.

(i) The dilution of different components of the tissue will depend upon the distribution of water, the diluent and the buffer, within each of the phases of the system at that particular temperature. For example, the cytoplasm with its salt and amino-acid composition will take up much more water than the lipid and protein parts of the cells; the same is true for sucrose. Indeed, readjustments of all the concentrations of the completely or partially soluble components of the tissue will occur in respect of a water-sucrose mixture, depending upon the activities of all components of the system, and the resultant solubility products relative to the sucrose solution in that condition; the relative permeabilities to molecules of all sizes of the parts of the tissue, both before and after homogenization, will also determine the net influx and efflux of water and sucrose. These movements will also be affected by the swelling (see opposite page).

The rates of each of the physico-chemical processes quoted above will depend on the temperature, which will be falling. The recovery of the tissue depends upon its 'rehabilitation' by incubation. Put another way, it depends upon how much the increase in entropy during preparation is reversed by incubation in optimal media. In general, the more alien the environment between the tissues existence *in vivo* and *in vitro*, the less likely it is to return to the *status quo ante*. We can, nevertheless, take refuge in examination of the 'functional' properties of the isolated tissue for comparison, where possible, with similar properties *in vivo*.

(ii) Chemical reaction of sucrose solution with tissues have long been assumed not to occur (see, for example, Pappius & Elliott, 1956); this will be considered in detail in the section on the effect of homogenization (see page 12).

The suspension of tissues in sucrose may have a preservative effect in that without all the co-factors in Krebs-Ringer solution, metabolism must be minimal and the tissues might be in 'suspended animation'. This state would imply that it was not exhausting the substrate. When the cations for various reactions were subsequently provided on incubation, the metabolic processes could restart. Such a line of thinking is speculative, but it would be well worthwhile comparing the effect of homogenization in sucrose, with homogenization in a balanced buffered oxygenated solution at 37°C on the final enzyme activity of the subcellular fractions as measured.

(iii) Swelling of tissue always seems to occur when it comes into contact with a medium, even if that medium is isotonic, oxygenated, buffered, and at 37°C. It is attributed to 'damage' of cutting or homogenization, and cannot be prevented even by hypertonic media. The 'damage' is said to be due to the cut cells 'dying', and their cytoplasm diffusing out. Hypoxia, high potassium, lack of substrate, or glutamate, have all been shown to increase the swelling of cerebral slices incubated *in vitro* (Pappius, Rosenfeld, Johnson, & Elliott, 1958; Joanny & Hillman, 1963). Perhaps we should entertain the suspicion that any substantial change in the tissue environment might do this.

One possible mechanism of swelling is presumably the breakdown by autolysis of large insoluble molecules into more, and smaller, active molecules. Alternatively, water could move into the tissue as a result of a Donnan equilibrium. Swelling is most easily measured by weighing the tissue before and after incubation, although it usually occurs immediately the tissue comes into contact with the medium, and very little more on incubation. Various methods have been used to measure whether the swelling is intracellular. They have normally been based on two assumptions, either that the marker – inulin, sucrose, or thiocyanate – does not enter the cells (Pappius & Elliott, 1956), or that the swelling is largely extracellular.

It is beyond the scope of the present study to consider this question in detail (see Hillman, 1966), but suffice it to say that the assumption that these 'markers' remain extracellular was originally without foundation and, when tested after many years use, has been found to be not true (Schousboe & Hertz, 1969).

We can only speculate on this question. It can be said with

certainty that if the swelling is intracellular, then all the concentrations within the cytoplasm and nucleoplasm would be decreased, at the same time the membranes would be stretched and their permeabilities might be altered. If indeed, the swelling is extracellular — and we must hope that it really is — the effect on the intracellular biochemistry would be minimal. If the swelling does not discriminate or is unequal, we have a real problem on our hands.

(iv) The tissue will lose many of its soluble components on coming into contact with any diluent. Usually, tissue is diluted prior to homogenization about one in twenty times; which will mean that in the absence of metabolism, there will be a net rapid efflux of cations, anions, amino-acids, monosaccharides, indeed all soluble small molecules. This will continue until real equilibrium has been reached between the tissue and the sucrose.

It was shown a long time ago that a liver nuclear fraction prepared in an aqueous medium, for example, could lose about half of its protein, and all its water soluble enzymes — adenosine deaminase and nucleoside-phosphorylase (Allfrey, Stern, Mirsky & Saetren, 1952). For this reason non-aqueous suspending fluids like cyclohexane-carbon tetrachloride have been used, sometimes with rapid freezing and lyophilization (Behrens, 1939; Dounce, Tischkoff, Barnett & Freer, 1950; Langendorff, Siebert, Lorenz, Hannover & Beyer, 1961). These preserved appreciable quantities of arginase, catalase, esterase and sodium ions. Organic solvents have the relatively minor disadvantage that they extract lipids, but this is a smaller price to pay than the loss of all water soluble components during homogenization; this realization seems to be spreading. Nevertheless, despite the long history of experiments unequivocally showing the disadvantages of aqueous media (for review, see Allfrey, 1959), they are still most widely used. It may be argued, and presumably will be, that 'by their fruits shall ye know them'; that is to say, if a subcellular fraction resulting from such treatment can perform oxidative phosphorylation, then that is good enough. This attitude will be dealt with in greater detail on page 105. Suffice it to say here that unless all the organelles *in vivo* have the same relative concentration of each of the water-soluble components — an implied belief as absurd as it is untrue — mixture of tissue with aqueous media would deplete them unequally. One could test for the presence of soluble constituents in the supernatant.

Tissues which are incubated subsequently in Krebs-Ringer solutions will regain much of their previous chemistry; for example, some of the potassium ions will be reconcentrated, the phosphocreatine will be re-synthesized, and the oxygen uptake will become steady. This indicates that the tissue is doing work against the medium. However, the current simpler incubating media, like Krebs-Ringer solution, do not contain amino-acids or soluble proteins, although media used for tissue culture must, e.g. '199' (Morgan, Morton & Parker, 1950). In higher concentrations, amino-acids may depolarize nerve cells (Krnjevic & Phillis, 1963) and cause loss of potassium ions from tissue (Joanny, Hillman and Corriol, 1966). Subcellular fractionation followed by subsequent incubation in what is considered an adequate biological medium further implies the assumption that each of the components will associate with the same concentrations of soluble materials as it had *in vivo* before homogenization and centrifugation. One's decision as to whether this assumption is warranted is a matter of personal opinion. Tissue for subcellular fraction is often finally incubated in buffered substrate mixtures, but rarely in balanced salt solution. The alternative is to demonstrate that in respect of the compounds finally measured in the subcellular fractions, initial dilution in sucrose produces the same final activity as initial dilution with a medium completely imitating extracellular fluid (see page 112). Presumably the reason why it is not is that pioneers in tissue fractionation found that 'pure' fractions of one type of subcellular organelle could not be obtained under these conditions.

The critical question about the tissue loss of soluble components is whether it will affect the enzymatic or other properties of the subcellular fraction. Obviously any reaction whose rate is dependent within the physiological range on the concentration of the compounds lost will be affected, unless they are replaced in the reaction mixture finally measured.

Allfrey, V. G. (1959) In *The Cell*, Vol. 1., Ed. by Brachet, J. & Mirsky, A. E., New York, Academic Press, page 193.
Allfrey, V. G., Stern, H., Mirsky, A. E., & Saetren, M. (1952) *J. Gen. Physiol.*, 35, 529.
Behrens, M. (1939) *Z. f. physiol. Chem.*, 27, 258.
Dounce, A. L., Tischkoff, G. H., Barnett, S. R., & Freer, R. M. (1950) *J. Gen. Physiol.*, 33, 629.
Hillman, H. (1966) *Int. Rev. Cytol.*, 20, 125.
Joanny, P., & Hillman, H. (1963) *J. Neurochem.*, 10, 655.

Joanny, P., Hillman, H., & Corriol, J. J. (1966) *J. Neurochem.*, **13**, 371.
Krnjevic, K., & Phillis, J. W. (1963) *J. Physiol.*, **165**, 274.
Langendorff, H., Siebert, G., Lorenz, I., Hannover, R., Beyer, R. (1961) *Biochem. Z.*, **335**, 273.
Morgan, J. F. Morton, H. J., & Parker, R. C. (1950) *Proc. Soc. Exp. Biol.*, (N.Y.) **73**, 1.
Pappius, H. M., & Elliott, K. A. C. (1956) *Canad. J. Biochem & Physiol.*, **34**, 1007.
Pappius, H. M., Rosenfeld, M., Johnson, D. M., & Elliott, K. A. C. (1958) *Canad. J. Biochem. Physiol.*, **36**, 217.
Schousboe, A., & Hertz, L. (1969) *Proc. 2nd Int. Mtg. of Int. Soc. for Neurochemistry*, Milan, Sept., 1969., page 354.

4. Homogenization of tissue

When tissue is homogenized, the following events occur:

(i) it is compressed;
(ii) the cells burst;
(iii) the intracellular and extracellular components are subjected to a new environment;
(iv) there is friction between tissue and tissue, tissue and suspending medium, tissue and homogenizer, and within the homogenizer itself.

(i) As the tissue is compressed, the pressure within the cells and their organelles must rise, while the cells' nuclear and mitochondrial membranes are completely or partially intact. The rate of flux across a membrane depends upon the pressure gradient on either side of it and, therefore, the flux of soluble materials across it will increase. The tighter the fit of the plunger to the homogenizing tube, the more closely it will approximate to a closed system. In such a closed system, the work causing the pressure rise will be translated into heat because the liquids are incompressible. The fate of this heat will be discussed in detail below, but it can be stated immediately that it is likely to cause a local temperature rise.

The quantity of heat generated will be different for each initially chemically distinct component or particles of the tissue, until the cells are burst. It will depend upon:

(a) the size of particle compressed, *e.g.* the cell, the nucleus, the mitochondrion;
(b) the shape of the particles;
(c) the total number of times the plunger rotates;

(d) the physical nature of each of the components mainly being subjected to pressure, *i.e.* the membranes; and
(e) the viscosity between the homogenate constituents, the suspending medium, and the homogenizer surfaces.

It would be desirable to measure the heat change on compression of cells but this would be very difficult. One would have to subject tissue to sufficient pressure so that its components would not quite fracture, and measure the temperature rise; this pressure would be different for each component. It will be seen on consideration of homogenization that even measurement of an overall temperature rise will not necessarily be useful since any change in enzyme activities finally measured in subcellular particles will depend upon heat generated *at the surface of those particles* and not on the overall temperature change of a tissue suspension, whose component heat conductivities and specific heats in any case are not known. Since four out of the five characteristics (a) to (e) upon which the heat generated depends, are not known, the heat generated cannot be calculated. Nevertheless, one could compress tissue in suspension under similar conditions to pressures just insufficient to fracture the cells, and examine the effect on enzyme activity. This would avoid the latter difficulties.

(ii) The cells burst due to the pressure of the homogenizer overcoming the forces joining the components together. Among these van der Waals forces are fairly weak and relatively little heat would be liberated because of them, although the heat that appears would be localized along the lines of the cracks. There does not appear to be any way of measuring this heat or the temperature rise resulting from it. Its magnitude would depend upon:

(a) the number of times the plunger was rotated;
(b) the fit of the plunger within the homogenizing tube;
(c) the physicochemical nature of the various membranes; and
(d) the shape and surface areas of the components of the homogenate at a given time.

(iii) As the tissue breaks up the intracellular organelles and extracellular materials become exposed to quite a new environment. The nucleoplasm, cytoplasm and extracellular fluids mix;

the nucleoplasm is high in DNA and probably sodium ion (Allfrey, Meudt, Hopkins & Mirsky, 1961; Lowenstein & Kanno, 1962; Itoh & Schwartz, 1957; Langendorf, Siebert, Lorenz, Hannover & Beyer, 1961), the cytoplasm is high in potassium ion and amino-acids (see Hodgkin, 1951; Lewis, 1952); the extracellular fluid is high in sodium ion and proteins. It has been suggested the rates of flux and biosynthesis of membrane materials *in vivo* are determined by the ionic gradients across the membranes at that particular instant (see Hillman, 1966). All these gradients and concentrations would be completely changed by the mixing of the different compartments and their considerable dilution with the suspending medium, say sucrose. Many trace compounds and co-factors might react with the sucrose solution on being diluted twenty times. This may well not matter, provided, firstly, that the sucrose is subsequently removed, secondly, that the preparation is incubated finally in a medium containing all trace compounds and, most important, that the effect of the sucrose solution is reversible. The magnitude of the effects of mixing will depend upon the relative concentrations of reactive substances within each original compartment.

Sucrose has been shown to inhibit many enzymes (Hinton, Burge & Hartman, 1969). This is dealt with in more detail in respect of sucrose gradients. Nevertheless, the amount by which it does should be measured and all measurements done in its presence should be corrected for it. Correlations between enzymes in biosynthetic pathways are particularly at risk without these corrections. Such remarks about sucrose apply equally forcibly to buffer, EDTA, bile salts or detergents, which are very often added to make fractionation easier. Their chemical reactivity is well known in other systems, and there would appear to be an urgent case too for investigating their 'unwanted' side reactions.

(iv) The friction occurring during homogenization will generate a considerable amount of heat. Since the dramatic experiments of Count Rumford in 1793, it has been known that friction generates heat. The amount generated will depend upon:

(a) the number of times that the plunger is rotated in the homogenizing tube;
(b) the coefficient of friction and the vicosity of each of the

pairs of reacting components; these are each dependent upon the temperature;
(c) the clearance between the plunger and the homogenizing tube;
(d) the total volume of the material being homogenized;
(e) the shapes, sizes and numbers of the particles involved;
(f) the area of the surface in contact at any given time; and
(g) continuing metabolism with the tissue. (b), (e) and (f) are not independent of each other.

It is undeniable that heat would be generated. By how much would the temperature rise? It is this which could alter the enzyme activity and protein configuration, hysteretically or irreversibly. This rise in temperature at the surface of a particular particle will depend upon:

(a) *the total heat generated* in its region, as above;
(b) the rate of heat generation;
(c) the heat conductivities of the tissue, sucrose and glass;
(d) the specific heat of each of the components including sucrose, glass, plastic, etc;
(e) the sizes, shapes, and packing of particles; and
(f) the ambient temperature in the region of the homogenizer.

These lists do not enable one to calculate heat generated or temperature rise, but they do help to identify the source of change which may affect the final answer. It also helps us to define some of the conditions which may decrease artefacts. From the two lists we may draw the following conclusions. It is desirable to homogenize the tissue the minimum number of times, the most slowly possible, in the coldest ambient temperature. At present only the latter consideration is recognized when the homogenization is performed 'ice cold'. The effect of the cooling of the homogenate is to make heat dissipation – as opposed to heat generation – as rapid as possible; it could minimize temperature rise if the heat conductivity of tissue were extremely good. Water conducts heat slowly enough – figures reviewed are between 12.7 and 16.5 x 10^{-4} cal. cm/sec/°C (Challoner & Powell, 1956); ice conducts even more slowly. Biological tissues conduct heat at about the same rate as water, published values being between 10 and 15×10^{-4} cal. cm/sec/°C (for review, see Spells, 1960; Hatfield, 1953). Very large temperature gradients may occur across tissue (Stone, 1938; Richter & Dawson, 1948).

Cooling tissue may actually *increase* the heat generated as it increases tissue viscosity, and the rise is non-linear with temperature for sucrose solution (Swindells, Snyder, Hardy & Golden, 1958). Thus, although heat dissipation may be accelerated, this effect may be completely counteracted by the increased friction generating greater heat. The real temperature change during homogenization would be almost impossible to measure as it is the temperature rise at the surface of the particles which is relevant to present considerations. At a molecular level temperature has no meaning, and the increased energy would be rapidly dissipated to surrounding molecules. Nevertheless, nuclei 10μ in diameter, or mitochondria 2μ in diameter, are several orders larger in size than would allow a mathematical treatment appropriate to molecules. In the author's opinion its irrelevant adduction has often led to fruitless disputes in relation to subcellular particles.

Homogenization is used so widely to 'release' enzymes, that it seems extraordinary that the other possibility, *viz.* that it might seriously alter the enzyme activity as measured, has only recently been entertained (Rodnight, Winter, Cook & Reeves, 1969). The latter group found, for example, that, whereas the percentage of protein in the 'microsomal' fraction remained constant at pestle speeds of $2\text{-}14 \times 10^2$ revolutions per minute, the 'mitochondrial' protein *rose* from 30 per cent to 55 per cent and the nuclear protein *fell* from 35 per cent to 20 per cent.

The practical reality of our ignorance of the values of each of the parameters upon which temperature rise during homogenization depends, can be circumvented by doing the following experiments: first of all, one can measure the bulk temperature rise following homogenization; secondly, one can homogenize the same tissue several times and measure the fall of its enzyme activity; one would then extrapolate back to zero homogenizations. This would give a correction factor. Real assessment of the temperature rise and quantity of heat generated in a particular system could be done by homogenizing a temperature-sensitive reaction mixture, calibrating its temperature sensitivity, and then measuring the specific heat of the system. A third experimental approach would be to isolate a particular fraction, measure its activity and then pass the fraction through the procedure again, and obtain an overall correction.

In an extremely significant study, Wiseman & Jones (1971) have compared the solubilization of α-glucosidase and invertase from yeast, by nine different commonly used disruption techniques. They found that particular disruption techniques gave quite different enzyme values (up to two orders), depending upon the apparent subcellular localization of their enzymes. It was also pointed out that the actual values did not necessarily correlate with the breakage seen on microscopic examination. These findings, if valid more generally, cast doubt on any comparisons of enzyme activities from animal tissues disrupted by different techniques.

Allfrey, V. G., Meudt, R., Hopkins, J. W., & Mirsky, A. E. (1961) *Proc. Nat. Acad. Sci. U.S.*, **47**, 907.
Challoner, A. R., & Powell, R. W. (1956) *Proc. Roy. Soc. B.*, **245**, 259.
Hatfield, H. S. (1953) *J. Physiol.*, **120**, 35P.
Hillman, H. (1966) *Int. Rev. Cytol.*, **20**, 125.
Hinton, R. H., Burge, M. L. B., & Hartman, G. C. (1969) *Analyt. Biochem.*, **29**, 248.
Hodgkin, A. L. (1951) *Biol. Revs.*, **26**, 339.
Itoh, S., & Schwartz, I. L. (1957) *Amer. J. Physiol.*, **188**, 490.
Langendorf, H., Siebert, G., Lorenz, I., Hannover, R., & Beyer, R. (1961) *Biochem. Z.*, **353**, 273.
Lewis, D. R. (1952) *Biochem. J.*, **52**, 320.
Lowenstein, W. R., & Kanno, Y. (1962) *Nature*, **195**, 462.
Richter, D., & Dawson, R. M. S. (1948) *Amer. J. Physiol.*, **154**, 73.
Rodnight, R., Wynter, C. V. A., Cook, C. N., & Reeves, R. (1969) *J. Neurochem.*, **16**, 1581.
Spells, K. E. (1960) *Physics in Biol. & Medicine*, **5**, 139.
Stone, W. E. (1938) *Biochem. J.*, **32**, 1908.
Swindells, J. F., Snyder, C. F., Hardy, R. C., Golden, P. E. (1958) *Supplement to National Bureau of Standards Circular*, **440**, Washington, U.S. Dept. of Commerce.
Wiseman, A., & Jones, P. R. (1971) *J. Appl. Chem. Biotechnol.*, **21**, 26.

5. Sucrose gradients

When the homogenized tissue is applied to a sucrose gradient, each part of the homogenate is subjected to:

(i) a different concentration of sucrose;
(ii) a different osmotic pressure, which will be a function of the water as well as the sucrose concentration;

(iii) a different viscosity;
(iv) a different amount of work during centrifugation; and
(v) a different error in the absorption measurement if optical methods are used.

(i) In a recent important piece of work, Hinton, Burge & Hartman (1969) have examined the effects of molar concentrations of sucrose on the following enzyme systems: alcohol dehydrogenase, AMPase, glucose-6-phosphatase, acid phosphatase, succinate-reductase, acid ribonuclease, and alkali ribonuclease. They found inhibitions by molar sucrose of 50, 65, 50, 20, 60, 20 and 20 per cent, respectively. The inhibition was largely reversible on dilution. Even greater inhibition may be achieved if concentrations of up to 2 molar are used (Hartman, G., personal communication). Although these authors found a considerable recovery in time, it cannot be stressed too strongly that each of the fractions in the gradients having been exposed to different sucrose concentrations will have been inhibited to different extents *in the same experiment*. Therefore, rapid separation and enzyme measurements, especially carried out without washing off the sucrose, are strongly contra-indicated.

(ii) It is not clear from the paper cited above whether sucrose inhibition is a 'specific' effect or an osmotic effect, or both, and this certainly needs urgent examination. It would be expected, in any case, that different osmotic pressures would engender relatively different equilibria of soluble materials between the particles and the suspending medium. The sucrose concentration has been shown to affect the density measurements of catecholamine granules, but not mitochondrial preparations (Lagercrantz, Pertoft & Stjärne, 1970). It also affects the specific activity of rat mammary mitochondrial pyruvate decarboxylase up to six times, and rabbit preparations up to twenty times (Gul & Dils, 1969).

(iii) Sucrose gradients may have linear concentrations, but the viscosity is not linear with concentration nor with temperature (see Swindells, Snyder, Hardy & Golden, 1958). In the cold, viscosity is increased and high speed centrifugation is always carried out in refrigerated conditions. Furthermore, in different layers of the centrifuge tube, different quantities of heat will be liberated (see page 21). This will cause different temperature rises, which will affect the viscosity and thus the heat generated.

These two factors will tend to cancel each other out; the higher viscosity due to higher concentration will cause more frictional heat which will warm the sucrose and decrease its viscosity.

(iv) The different amount of heat generated during centrifugation will be dealt with under the latter heading (page 20). Suffice it to say here that if the rate of heat conduction away from each layer were different, one would certainly find different enzymic activities on measurement, *even if each layer originally had the same activity*. In saying this the assumption is that the heat generated is relatively large. In the absence of measurement or reasonable mathematical calculation, it may be guessed as being either insignificant or enormous. There is no alternative to measurement; techniques for assessing this are suggested in the section on centrifugation (see below).

(v) There will be different errors in measurement of absorption of light through sucrose depending upon:

(a) the size and shape of the particles in each fraction;
(b) the absorption of light by tissue due to 'interfering' species other than the one being measured, even at chosen wavelengths;
(c) the considerable absorbance of nucleic acids at 260 mμ wavelength compared with other substances present; and
(d) the absorption of light by sucrose; this is relatively unimportant.

Gul, B., & Dils, R. (1969) *Biochem. J.*, 111, 263.
Hinton, R. H., Burge, M. L. E., & Hartman, G. C. *Analyt. Biochem.* 29, 248.
Lagercrantz, H., Pertoft, H., & Stjärne, L. (1970) *Acta Physiol. Scand.*, 78, 561.
Swindells, J. F., Snyder, C. F., Hardy, R. C., Golden, P. E. (1958) *Supplement to National Bureau of Standards Circular,* **440**, Washington, U.S. Dept. of Commerce.

6. Centrifugation

This consists of placing the homogenate suspension in a plastic, metal, or glass tube of a given length, at an angle to the vertical axis, in a refrigerated centrifuge. The tube containing the suspension, usually in a vacuum, is accelerated to a given speed, held at that speed for a prescribed time, and then decelerated passively, or by a brake. The different constituents, having different densities and shapes, separate into layers. The following events occur during this operation:

(i) some components of the tissue shear;
(ii) the particles move due to centrifugal forces;
(iii) the homogenate suspension is subjected to high pressure;
(iv) the particles travel into an *increasing* and high pressure gradient at different rates;
(v) some particles impact at the bottom of the centrifuge tube;
(vi) the centrifuge tube is being stretched adiabatically;
(vii) the centrifuge bearings are generating heat;
(viii) there is aerodynamic heating; and
(ix) the refrigeration is cooling the system.

(i) The effect of shearing may be to separate different components, *e.g.* nuclei from cytoplasm, cell membranes from cytoplasm, etc. This will probably involve the breakage of bonds formed by van der Waals forces. That shearing does occur is evident from the much greater ease of separating red cells from plasma than nuclei from mitochondria.

(ii) The particles move down the column of suspending medium through the centrifuge tube. Friction occurs and the amount of heat generated will depend upon:

(a) the coefficients of friction between each of the different kinds of particles and each other, the suspending medium and the side of the tube. Each of these viscosities will be unique for a particular interface and will depend on the specific gravity, the chemistry, the shape, the temperature, and the pressure. Although theoretically they could be measured, their value operationally probably could not be calculated or measured without making several unwarrantable and untestable assumptions about physical orientations and chemical states within the system;

(b) the density, size, shape, and packing of the particles at any time. These will affect the friction between them, but will also determine the centrifugal force acting upon them. They will themselves be determined by the pressure, temperature and chemical environment, within the part of the tubes through which they are travelling;

(c) the centrifugal force applied; this depends upon the number of revolutions per minute of the tubes — whether they are accelerating or decelerating; of course even when they are travelling at constant speed in a circular motion, they are accelerating; it also depends upon the exact distance of that particular fraction from the axis of rotation at that particular time; this determinant is usually ignored by calculating the centrifugal force at the midpoint of the centrifuge tube, as if the centrifugal force were equal throughout the tube, which it cannot be. This in itself invalidates any statement about the g force in respect of the isolation of a particular fraction;

(d) the position within the tube at which the particles have arrived *by a particular time*, until they reach their final layers, or the bottom of the tube. The heat generated by centrifugation is a function of the distance of the centrifugate from the axis of rotation. The centrifuge tube is at an angle to the vertical, and one can very rapidly calculate that in most centrifuges the bottom of the tube is subjected to a centrifugal force about twice that to which the top is subjected. If all other parameters were equal — which in gradient centrifugation they are definitely not — then twice as much energy would be injected into the system at the bottom of the tube, and its temperature rise would be twice as great. Any temperature sensitive system which was hysteretic would be changed more at the bottom of the tube than at the top. Once again, one can see that centrifugation under these circumstances only *once* could well induce different degrees of change within a single tube, which might be reflected in different enzyme activities measured subsequently, *even if* all fractions originally had the same activity. This real difficulty is circumvented verbally by measuring the average centrifugal force in the middle of the tube; this is irrelevant with respect to a particular fraction, for which the centrifugal force to which it has been subjected should be specified. Perhaps the centrifugal forces should be stated at particular times until the fraction comes to its final position, and these forces should be integrated with respect to time; the final centrifugal force acting at that particular layer for the length of time during which it rests there should also be given. In this respect then, it might well be more appropriate to calculate the centrifugal *work* done on the fraction. This will be proportional to the heat generated, but not necessarily the temperature rise (*q.v.*).

It is clear from this short discussion that the usual quotations of centrifugal force applied in subcellular fractions are more in the nature of operating instructions for the instrument users, rather than statements of work done in isolating each fraction.

The effect of rotor velocity on the measurement of sedimentation coefficient of nitrocellulose was studied in a paper from Svedberg's laboratory (Mosimann & Signer, 1944). They calculated how the pressure induced at different speeds would affect the viscosity and specific gravity, and they produced a table of corrections. More recently, Polson (1967) found that sedimentation coefficients of ovalbumin and human γ-globulin were 10 per cent higher at rotor velocities below 10,000 rev/min than at 56,100 rev/min; haemocyanin of *Jasus lalandii* showed the same effect. As is pointed out in the latter paper, it has usually been assumed that calculated sedimentation coefficients are independent of rotor velocities.

(iii) The following is the equation for the radial pressure distribution in a rotating liquid (see Appendix 2):

$$p = \frac{\rho \omega^2 r^2 + C}{2}$$

where ρ is the density of the suspending fluid;
 ω is the angular velocity;
 r is the radius of rotation; and
 C is a constant, equal to twice the pressure within the centrifuge.

Unfortunately we do not know the *real* density during centrifugation, but we can immediately see that if C is small the pressure at a particular point in the tube is a function of the square of the radius of rotation; this would give a pressure ratio of 3 to 4 between the bottom and the top of most centrifuge tubes. Like the density, the constant is unknown, and might well be unique for the different layers being centrifuged.

(iv) The pressure at the bottom of a homogenizing tube is given by the force exerted by the column of fluid above it. If one uses a value of approximately 1.0 for the specific gravity of the homogenate and suspending medium, and considers a centrifuge tube 10 cm in length, one can calculate that each 1000 g centrifugal force will induce a pressure at the bottom of the tube of about 1 atmosphere. Such a pressure has been shown to *increase* the activity of pepsin, trypsin and diastase (Frankel & Meldolesi, 1921); 1500 atmospheres partially inhibits pepsin digestion of gelatin, but reverses on removal of pressure (Benthaus, 1942). Much higher pressures denature proteins and inhibit enzymes (see Table 2). Although in a fluid, much of the pressure will be translated into internal energy, the resultant temperature rise can be shown to be small (see Appendix 3).

Some new adventitious reactions between tissue components may occur during centrifugation. The effect of pressure on the *rate* of these reactions is given (Ingraham, 1962) by the equation:

$$V_p = V_o \exp\left(\frac{-p\Delta VI}{RT}\right)$$

where V_p is the rate of the reaction at pressure p;
V_o is the rate at the lower pressure;
p is the pressure; and
ΔVI is the change in volume between the normal and the activated states.

It will be seen that even if there is a minimal change in volume of the suspending medium — as that is a fluid — the rate of reaction will be a logarithmic function of the change in volume of the suspended particles. This will depend upon their elasticity during centrifugation, about which we have no data.

If we were to assume that the pressure change were adiabatic and reversible, we could derive an equation for the temperature change in a pure substance. This would indicate what were the relevant parameters. Such an equation is derived in Appendix 3.

$$\frac{T_2}{T_1} = \exp\left(\frac{\beta}{\rho C_p}\Delta p\right)$$

where T_1 and T_2 are the initial temperature and that resulting from pressure, respectively;
    β is the volumetric coefficient of expansion;
    ρ is the density of the solution under those conditions;
    C_p is the constant pressure specific heat; and
    Δp is the change in pressure.

Unfortunately, the volumetric coefficient of expansion is not known at the temperature of $-10°C$ at which the centrifuge is alleged to be, or at the *real* temperature within particular fractions of the homogenate; neither is the density of the solution, nor the constant pressure specific heat. Although some of these values have been measured in constant, static, and simplified systems, we cannot adduce these measurements, due to lack of data about the real physico-chemical environment of the tissue.

In general, we can say that the chemical potential – and therefore reactivity – of *each* of the many species will be dependent upon the pressure, the temperature and their partial molar volumes in respect of the suspending medium (for discussion, see Spanner, 1964). Of course, the suspending medium and each of the components of the tissue must be considered as a different phase. Unfortunately, here again, we have no data about the magnitude of these parameters during centrifugation. We could, nevertheless, signify it as appropriate to examine the effects of the comparatively low pressures incidental to centrifugation on the reactivity, enzymic activity, and proteins in centrifugates.

All that has been said about the pressure occurring during centrifugation is also true for homogenization (see page 12), although the tissue is subjected to pressure during homogenization for a much shorter time. All authors agree that neither activation nor inhibition are linear with pressure, and that both these effects, and their reversibility, are pH sensitive. Sometimes the intensity of the effect depends upon the rate of application, and sometimes it is independent of it.

All this clearly dictates the necessity to calculate the possible pressures induced during centrifugation, which will be different at different points within the centrifuge tubes. With the increasingly

Preparation	Pressure (bars)	Effect after treatment	Reference
Egg albumen	7,000	Complete coagulation	Bridgman (1914)
Serum albumen	4,000	Reversibly denatured	Suzuki, et al. (1963)
Tetanus toxin	13,500	Complete loss of activity	Basset & Macheboeuf (1932)
Tuberculin	13,500	No change	
Yeast invertase	13,500	100% inhibition	
Cobra venom	13,500	No change	
Pancreatic juice	13,500	Inhibits lipase; no effect on trypsinogen or amylase	Basset, et al. (1933)
Pepsin	6,000	100% inhibition	Mathews, et al. (1940)
Rennin	6,000	100% inhibition	
Trypsin (crystalline)	7,600	81% inhibition	Curl & Jansen (1950a)
Chymotrypsin (crystalline)	4,800	63% inhibition	
Pepsin (crystalline)	6,100	100% inhibition	Curl & Jansen (1950b)
Chymotrypsinogen (crystalline)	5,500	100% inhibition	

Table 2 The effects of pressure on proteins and enzymes. The activities were tested *after* the substances had been subject to pressure, and were mostly measured at room temperature.

powerful instruments coming into routine use in laboratories, one has to bear in mind the effects, a few examples of which are shown in Table 2. However, calculations made on assumptions about parameters will never be a satisfactory substitute for doing serious control experiments on the effects of centrifugation on the system one is studying (see below).

Relatively little work has been done on the mechanism of the effects of pressure, though the resultant denaturation has been reviewed as early as 1954 (Johnson, Eyring & Polissar). Most studies have examined the effect of denaturation on volume change (Kauzmann, 1959; Gill & Golgovsky, 1965; Brandts, 1969), though these give important insights into pressure effects on protein conformation. An excellent recent review mostly deals with effects on unicellular organisms, (Zimmerman, 1970).

The particles move into an increasing pressure gradient. This will be a region of higher viscosity, higher sucrose concentration, higher particle crowding, and higher centrifugal field; these will have the effects of increased heat generation, increased enzyme inhibition, increased pressure, and increased gravity, respectively. Such different factors will interact; for example, the greater heat generated due to higher viscosity will induce greater enzyme activity, but the higher sucrose concentration will inhibit it; increased pressure will make the particles smaller, so that less friction will occur with the medium.

In order to evaluate the work done on a particle moving in a centrifugal field, one would have to determine the following forces; firstly, the viscous drag at the boundary between the particle and the suspending medium; secondly, the hydrodynamic force present as a result of the velocity gradient across the centrifuge tube, and, thirdly, the net inward thrust produced by the radial pressure distribution. Investigation of the relative magnitude of these forces using realistic parameters would be useful.

(v) Some particles impact at the bottom of the tube. A certain amount of energy is imparted to the particles by centrifugation. Some of it will be dissipated as heat on its journey down the tube, and some will be lost on impact at the bottom; this latter quantity may be small, and in any case, it will be released very slowly.

Assuming that the particles lose all their potential energy, their kinetic energy at the time of impact will be $\frac{1}{2}mV^2$, where m is their mass, and V their velocity. Usually, a centrifugation run is carried out with the intention that only one fraction arrives in the pellet. Therefore, this material, and only this material, will be subjected to the heat liberated by the impact. Here again, we have another reason why different amounts of heat will be generated in the different fractions in an apparently single procedure. The temperature rise will be determined by several other factors (see page 15).

(vi) The system is deformed, and strain is induced in the centrifuge tube, its container, and the rotor. The stretching will cause loss of heat from each of these components, which will re-warm on relaxation. If in the present system, the stress-strain relationship is linear, then the strain energy equals:

$$\frac{0.5 P^2 L}{AE}$$

where P is the applied stress;
L is the length before loading;
A is the cross sectional area; and
E is the modulus of elasticity (Roark, 1954).

The load is equal to the mass of each of the components multiplied by the number of g of centrifugation. The strain energy is thus a function of the square of the centrifugal force. All the stress of the centrifugation will be translated into strain within the containers, made of plastics, aluminium and titanium. This situation is one of the few in which the amount of heat generated could be calculated reasonably accurately, but the dissipation of heat away would depend upon many other factors, which would be difficult to measure. Thus the temperature rise and its effect on the centrifugate cannot be assessed, but may be small.

(vii) The centrifuge bearings and the rotor generate heat however cunningly the engineers design them. The intention is that the minimum friction be induced in these regions, that the heat be allowed to dissipate quickly by the use of conductive materials, and that the refrigeration unit be as powerful as possible. The heat dissipation takes place mostly through the metal spindle and rotor, as the centrifuge is normally evacuated of air to reduce aerodynamic heating. Thus the maximum heat dissipation is limited by the real rate at which the piece of metal of a particular shape can conduct it away.

The amount of heat generated will depend upon the speed of the rotor, the wear on its bearings, the material of the rotor, its unique shape, whether the centrifuge is accelerating or decelerating, whether it is being braked or not, the real coefficient of friction and the duration of the centrifugation. Naturally, the temperature rise, also, will depend upon all these parameters, as well as the heat conductivities of all the components, and the degree of vacuum induced within the centrifuge chamber.

In a recent survey of four well-known companies selling high speed centrifuges, the author was informed that the temperature rise measured somewhere in the region of the rotors did not exceed 4°C. This was reassuring to some extent viewed as sales information, but none of the manufacturers could give details of the temperature rise *within the homogenizing tube*, arising from the bearings. D. Fargie, R. V. Read and H. Hillman (1966, unpublished) attempted to measure this without success. The difficulty was that the centrifuge must be refrigerated while it is in use and during its slowing down. Stopping it takes a long enough time for the temperature to have fallen again during deceleration. In order to measure the temperature rise during the centrifugation itself, one should take leads *from* a temperature sensing device within the homogenizing tube; these *leads* would generate their own friction. A thermistor to 'broadcast' the heat change would be separated into its own 'subcellular' components. A further approach was to centrifuge a temperature sensitive protein. Centrifugation of 1 in 20 egg albumen for 5 minutes at 1000 rev/min on the dial resulted in significant change of ultraviolet absorption.

These experiments did not permit the interpretation that the rotor was the only source of heat, or 'denaturation'. Nearly all the other sources of heat mentioned in this section could have contributed. Nevertheless, it would be highly advantageous if a manufacturer could produce a centrifuge in which the temperature could be monitored within or near the homogenate – preferably throughout the experiment.

(viii) Aerodynamic heating will result from friction between the homogenizing tubes and rotor, on the one hand, and the residual gas left in the bowl, on the other. The gas will include air, water vapour, and other gases diffusing from the sample, the plastic cups, the rubber stops, the bearings, etc. The pressure is usually down to a few mm of mercury, and so this would probably not be a large source of heat. However, its magnitude would depend upon the efficiency of the pump, the efficiency of the refrigeration unit, the actual speed and acceleration of the centrifuge, leaks within the system, and the materials of the centrifuge. It would not be expected to be great. Of course, the higher the vacuum, the less the aerodynamic heating, but the more effectively the system is isolated. Thus heat generated within the centrifuge tubes, within their walls, and within the rotor, will not be easily dissipated. It will be lost from the tubes mainly through radiation, and much of this will be conserved by total internal reflection within the smooth shiny metal surfaces of the centrifuge bowl, as it is in a thermos flask.

(ix) Refrigeration is normally used to increase the rate of heat dissipation. Its effectiveness depends upon the power of the refrigeration plant, and the size and total geometry of the inside of the centrifuge. The centrifuge and the rotor are usually pre-cooled before use, and the refrigeration is continued during the stopping of the centrifugation.

However, it must be insisted that no amount of refrigeration will decrease the amount of heat generated within the homogenizing tube. Indeed, increasing the preparation's viscosity may increase it (see page 18).

Basset, J., & Macheboeuf, M. A. (1932) *Compt. Rend. Acad. Sci. Paris,* **195**, 1431.
Basset, J., Lisbonne, M., & Macheboeuf, M. A. (1933) *Compt. Rend. Acad. Sci. Paris,* **196**, 1540.
Benthaus, J. (1942) *Biochem 2.,* **311**, 108.
Brandts, J. F. (1969) in *Structure and Stability of Biological Macromolecules*, ed. Timasheff, S. N., & Fasman, G. D., New York, Marcel Dekker, page 213.
Bridgman, P. W. (1914) *J. Biol. Chem.,* **19**, 511.
Curl, A. L., & Jansen, E. F. (1950a) *J. Biol. Chem.,* **184**, 45.
Curl, A. L., & Jansen, E. F. (1950b) *J. Biol. Chem.,* **185**, 713.
Frankel, S., & Meldolesi, G. (1921) *Biochem. Z.,* **115**, 85.
Gill, S. J., & Golgovsky, R. L. (1965) *J. Phys. Chem.,* **69**, 1515.
Ingraham, J. L. (1962) in *Bacteria. Vol. IV Physiology of Growth,* ed. Gunsalus, I. C. Stanier, R. Y. New York, Academic Press, page 268.
Johnson, F. H., Eyring, H., & Polissar, M. J. (1954) *The Kinetic Basis of Molecular Biology*, New York, Wiley.
Kauzmann, W. (1959) *Adv. Protein Chem* , **14**, 1
Mathews, J. E., Dow, R. B., & Anderson, A. K. (1940) *J. Biol. Chem.,* **135**, 697.
Mossiman, H., & Signer, R. (1944) *Helv. Chim. Acta.,* **27**, 1123.
Polson, A. (1967) *Biochem. J.,* **104**, 410.
Roark, R. J. (1954) *Formulas for Stress and Strain,* New York, McGraw Hill, page 75.

Spanner, D. C. (1964) *Introduction to Thermodynamics,* London, Academic Press, page 152.
Suzuki, K., Miyosoura, Y., & Suzuki, C. (1963) *Arch. Biochem Biophys.,* **101**, 225.
Zimmerman, A. M. (1970) Ed. *High Pressure Effects on Cellular Processes*, New York, Academic Press.

7. Addition of substrate mixture

This addition to the centrifuged fraction will cause:

(i) dilution of the tissue, including the enzyme;
(ii) dilution of the layers in the different sucrose concentrations; and
(iii) partition of the substrate between different fractions.

(i) The degree of dilution will obviously depend upon the relative quantities of the fraction and the substrate mixture, but the dilution of the enzyme will depend upon its partition coefficient between the particles, the sucrose, and the substrate mixture; the distribution of enzyme and its activating co-factors might be altered when the third of these constituents is added.

(ii) The different layers of the sucrose gradient if diluted equally by the same volume of substrate mixture will inhibit the enzyme activities unequally, as the reversibility of the sucrose effect depends upon washing it out completely (see discussion on page 18).

(iii) Partition of substrate between different fractions does not depend only on the affinity between the pure substrate and the pure enzyme. It is a function of the very complex chemistry of all the other constituents of the fraction as well. If one is claiming that a particular fraction has an enzyme activity relative to another fraction, this has two possible meanings. The usual one — which we may call the operational meaning — is that if one adds the same excess quantity of substrate to all the fractions, one can measure and compare the subsequent change of substrate concentration after incubation with each fraction. It would be more relevant to find out the affinity of the substrate for each of the fractions first, and then correct the enzymic activities for these. On the other hand, if one is claiming the presence of an enzymic activity in a preparation, one must also demonstrate that the preparation has negligible non-enzymic effect — or measure how much effect there is; boiled or inactivated preparations can be used as controls, and values should always be corrected with them (see page 39).

8. Extraction of the products of the enzymic reaction

When one is extracting products from a number of different fractions, the proportion of each product extracted cannot be assumed to be similar. It will depend upon:

(i) any alterations in the reactivity of the fractions due to the extractant;
(ii) the affinity of the extracting agent for each of the different chemicals in the heterogeneous fraction;
(iii) the heats of reaction of the extractant with each of the components;
(iv) the chemical affinity of the enzyme for each of the components of the system; and
(v) the chemical affinity of the substrate for each of the components of the system.

When the extracting reagents are added, they will each dilute each of the components. The extracting agents may be strong acids or alkalis. They will react with the components of the tissue, and generate a great deal of heat in many cases, even if they are subsequently cooled. One need only to think of the heat liberated on dilution of sulphuric acid with water. This heat, accompanied by the massive change in pH and the arrival in the vicinity of large quantities of foreign anions as well as the hydrogen ions, will induce new chemical reactions between the components which would not have occurred under natural conditions. For this reason, a simple chemical system without extraction is proposed (see page 111).

All of these functions are extraordinarily complex, as they will depend upon the chemical nature of all the active molecules present, their physical states with respect to temperature and concentration, their entropy state at the time of extraction, and many other factors, mostly unknown, and many unknowable. Nor do we have an equilibrium system here, as it is a common experience that the activity of an enzyme extracted after centrifugation is a function of the rapidity with which it is measured after preparation.

Denaturation of proteins during centrifugation may affect their affinity for substrates or extractants. Denaturation has been shown to alter the partition coefficients in a dextranpolyethylene glycol system very considerably (Albertsson & Nyns, 1961), and DNA in the same system may alter its distribution 1000-fold (Albertsson, 1962).

Before any comparison of subcellular enzyme activity can be made, the following examinations must be done: (a) the location of added enzyme after mixing with each subcellular fraction; (b) the affinity of added substrate for each fraction; and (c) the extractability of the products of enzyme activity. It is only too well known that proteins can 'interfere' with flame photometry, amino-acid measurements, lipid extractions etc., yet enzyme activities are often naïvely compared with the mainly aqueous supernatant solution, and the protein and lipid 'mitochondrial' fraction, without previous recovery measurements having been made.

The difficulty of varying extractabilities is minimized by using optical methods, but they have their own difficulties, due to other materials affecting the absorption of light.

9. Measurement of product of the reaction

This subject is such a wide one that few precise comments can be made. Colorimetric, fluorimetric, radioactive, and many other methods are used to measure the final product. They share two difficult but not insurmountable problems.

In order to measure one macromolecule in a mixture, a compound, or an organelle, one may have to extract it. This implies perforce that one has changed its chemical activity; that is, that the activity of the native material *in vivo* is different than in extract. Ideally we should set ourselves the Herculean task of trying to measure chemical activities *in vivo*. Failing this, we should perhaps try to measure the activities more often in an environment simulating the extracellular and intracellular fluid, *i.e.* at the relevant pH, ionic concentrations, oxygen tensions, etc. Enzyme essays seeking to characterize 'optimum' — that is maximum — activity, however achieved, may not yield useful information about biochemistry *in vivo* as characterized (see page 111).

Since substances normally associated with the ones to be

assayed may change their measured activities, it is important that all instruments be calibrated *not* with pure solutions, but by adding known quantities of the material to be measured to the system in which it is normally found, and examining recoveries. This is not done very often. Obviously, if it were to give a different answer than the more usual calibrations with pure solutions — as is likely — the latter techniques should be abandoned.

Optical methods used for examining reactions as they are going on, without extraction, have obvious advantages, but they make a gross assumption, that is, that chemical changes in the solutions other than those which we wish to measure do not alter the optical properties of the ones in which we are interested. This may or may not be true.

Discussion of optical techniques is beyond the scope of the present study, but they have been dealt with in considerable detail (see for example, Oster & Pollister, 1956; Udenfriend, 1962).

Assumptions inherent in techniques attempting to examine subcellular localization of enzymes

Many of the assumptions implied in the use of subcellular fractionation have been discussed with respect to particular steps in the process, but a few further remarks are appropriate after listing the main assumptions.

(a) That killing the animal has no substantial effect on its biochemistry.†
(b) That cooling does not induce substantial irreversible change in the isolated tissue.
(c) That the enzyme activity of a homogenate decreases linearly with dilution.
(d) That the medium in which the tissue is homogenized, *e.g.* sucrose — or additives which make organelle separation easier, *e.g.* ethylene diamine tetracetic or bile salts — do not alter the chemical activity significantly and irreversibly.†

† In some experimental systems this has been shown to be untrue.

- (e) That the enzyme activity measured finally is not significantly changed by the incomplete replacement of soluble constituents, like cations or amino-acids, which are lost on gross dilution, homogenization and centrifugation of tissue in a different environment.
- (f) That movement during preparation of known co-factors, like calcium or magnesium, or unknown ones, will not alter substantially the apparent location of enzyme activity as measured.*
- (g) That soluble materials originating from *any* part of the cells will not diffuse into the supernatant during preparation, and thus be supposed to have originated in the cytoplasm.*
- (h) That the heat generated during homogenization is so rapidly conducted away that the temperature does not rise sufficiently high to produce significant irreversible change in enzymes.
- (i) That refrigerating the centrifuge prevents temperature change at the particle surface.*
- (j) That enzyme activities are not irreversibly changed by pressure.†
- (k) That the same amount of work is done on the homogenate in different parts of the centrifuge tube.*
- (l) That the same amount of heat will be generated in parts of the homogenate with differing viscosities.*
- (m) That the extraction from each of the final fractions is equal and complete.†
- (n) That the enzyme preparation has no significant non-enzymic effect on the substrate.
- (o) That the similarity of the appearances on electron microscopy of the parent tissue and its subcellular particles shows that its biochemical properties have not been changed by the fractionation.

The assumption (c) — that enzyme activity decreases linearly with dilution — is true if excess substrate is provided for the reaction; *in vivo*, this might well not be so. When one dilutes

---

* Assumption contradicts laws of thermodynamics or physics.
† In some experimental systems this has been shown to be untrue.

tissue, one dilutes a multicomponent system whose components may not react linearly. It can be shown that normally this does not matter in the presence of excess substrate, but one is also diluting co-factors, whose ability to activate a system may not be linear with their concentration. This assumption should be tested with pure enzymes, with pure enzymes added to tissue, as well as with endogenous enzymes.

Additives to a preparation which permit easier particle separation or enzyme extraction, like EDTA or bile salts, will have an effect on enzyme activity. For example, any of the 27 out of the 29 enzymes listed as being divalent-ion activated or inhibited (Dixon and Webb, 1962) would be affected by excess EDTA addition at any stage of the procedure. It is desirable to test if the effect would be reversed on subsequent addition of the co-factors. However, although calcium and magnesium ions are often added in substrate mixtures, manganese, cobalt, iron, nickel and zinc are not; they may, however, be present as trace metals in large enough concentrations in the other components of the mixture.

Natural detergents, like bile salts, and synthetic ones, like Triton or Teepol, are used to break up mitochondrial membranes and 'release' their enzymes (for review, see Morton, 1955); they are, therefore, very powerful chemical agents. Koch and Lindall (1966) have shown that 0.5 per cent deoxycholate activates cytochrome oxidase from rat brain fractions, three to twenty-five times; while Moore and Lindall (1970) found up to seventeen times as much activation for the same enzyme, and one to eight times for $NADH_2$-cytochrome reductase; much more worrying than the activation *per se* was the fact that each fraction showed different activation.

Assumption (f) also related mainly to movements of co-factors. If, for example, the lipid of the so-called mitochondrial fraction has a great affinity for calcium ions, any enzyme which is calcium activated (*e.g.* phosphatase) will appear as a higher concentration in that fraction. This will not necessarily depend upon where the divalent cation was *in vivo*, but where it arrived after its complicated journey during the preparative procedures. There is some evidence that cut surfaces of cerebral tissue accumulate calcium which was not there before (Lolley, 1963). The basic difficulty is that the measurement of enzyme activity instead of enzyme concentration implies that the activity of the preparation is not altered by anything else in the system. This will be discussed further (see page 117, (i)).

The assumption that any enzyme found in the supernatant necessarily originates from the cytoplasm or 'cell-sap' is contrary to the Second Law of Thermodynamics. If an enzyme, co-factor or any other component, is water soluble and can cross the homogenized membrane, and it is present *in vivo* in a higher concentration in one part of a cell, it will diffuse out into the supernatant along its electrochemical gradient. An equilibrium will occur between the tissue and the supernatant depending upon the particular partition coefficient of that substance relative to each chemical in its locality. Since water soluble enzymes are hydrophilic, a higher concentration would be detected in the supernatant than would be expected from the total preparation.

It would seem superfluous to state that a technique based on assumptions contrary to laws of physics or thermodynamics cannot but yield untrue information.

In respect of assumption (h), it is doubtful if anyone would deny that heat was generated during homogenization. Therefore it has to be asserted that either the heat is dissipated so rapidly that the temperature will not rise, or that the temperature rise is too small to affect the enzyme activity irreversibly. This problem has already been considered (see page 16). It would be relatively easy to test the effect of homogenization on enzyme activity, and then correct for it.

Here we peer into the depths of uncertainty, since it is usually accepted that the effect of homogenization is to break the cell walls and thus release the large molecules from within it. There is, however, another real possibility. That is that the heat generated during homogenization, or the new environment of the enzyme created by it, actually itself *increases* the measured enzyme activity. The experiment suggested on page 16 of homogenizing the same tissue several times should be able to distinguish between these two possibilities.

The usual belief is that the system really does approximate to an open system in such a way that the heat is dissipated rapidly. In respect of assumption (i), it must be stressed that refrigerating the tubes during homogenization or centrifugation will not affect the heat generated, except possibly to increase it (see page 16). One can strike a match in the Arctic winter because the heat is generated locally. The match will cool down more rapidly after the flame has burnt out. There is a peculiar naïvety in 'refrigerating' tissues subjected to homogenization, centrifugation, electrophoresis or sonication.

The effect of these treatments will always depend on the temperature rise, and thus the more slowly they are carried out, the less they will be changed. It will also depend on the reversibility of any temperature effect on the enzyme.

Assumption (i) is contrary to fundamental laws. The usual defence can be based only on the demonstration that the changes are so small that they may be ignored. Experiments to test this are painfully and embarrassingly overdue, especially if the effect should prove to be large. It would seem that the onus of proof here lies heavily on the believers.

Assumption (j) — that enzyme activities are unchanged by pressure during homogenization and centrifugation — has been examined (page 25, Table 2). When work is done compressing relatively inelastic liquids, there will be a considerable exchange of heat during its compression and relaxation, *i.e.* during acceleration and deceleration of the centrifuge to its intended maximum speed. Most authors cited (page 25) find a non-linearity about the pressure effects on enzyme activities, with a fairly distinct shoulder on the curve of pressure against activity between 5,000 and 7,000 bars. There has been an increasing tendency in recent years to use greater and greater centrifugal forces, generating pressures well beyond this.

Assumptions (k) and (l) are interdependent. The centrifugal force depends upon the distance from the rotor which varies in different parts of the centrifuge tube. The density of the sucrose gradient at a particular layer will also affect the centrifugal force, and thus the pressure. Both the pressure and the work done on different layers will be discontinuous functions. Assumption (k) is contrary to simple physical laws.

The amount of work done on the different layers will be the main determinant of the temperature rise but this will also be affected by several other parameters of the system, of which heat conductivities of its different components are the most important (see page 14).

It is a commonplace that extraction agents extract unequal quantities of solutes from different mixtures. The simple assumption

(m) is contrary to everyday experience and hardly merits further consideration.

Researchers rarely examine nowadays whether or not enzyme preparations have significant spontaneous activity – assumption (n) – in the way that the pioneers of enzymology often did. Very frequently the control samples in a preparation contain the substrate mixture and the co-factors, but no inactivated enzyme preparation. If an enzyme co-factor is being studied, it is usually assumed that the substrate will not be affected by the presence of the co-factors. Greater attention should be paid to the study of the properties of well-known substrates in the absence of tissue. It would be invidious to cite particular examples of the former two practices, since they are universal. However, if one teaches students that an experiment is only as good as its controls, one cannot abjure the moral obligation of carrying out such appropriate controls for all experiments in which it is being claimed that findings are in any way meaningful. Interpretations depend on what one really means by 'enzyme activity' (see page 117) and what control experiments are possible (see opposite page).

One of the commonest implied assumptions, (o), that a clear electron microscopic picture of a subcellular fraction is cogent evidence of the biochemical health of the subcellular particles, is dangerously misleading. The fractions have already been subjected to the whole procedure examined in this section; at this stage biochemical studies are carried out on them. One then proceeds to a further complicated series of steps, including 'fixation', dehydration, deposition of osmic acid salts, replacement of tissue fluids by Araldite or acrylic, and subjection of the osmic acid deposit to a very energetic beam of electrons. This technique will be examined in detail later (page 54). Suffice it to note here that it would be very surprising if the addition of all these powerful chemical reagents were not to alter the biochemical properties of the tissue. The comparison of the appearance of cellular organelles in a thin slice of incubated tissue with similar structures in subcellular fractions, can only indicate the very crudest similarity between the electron beam's photo of the osmic acid-reacting portions of the cells in each case.

It must be insisted that the validity of any localization of enzyme activity within a cell is dependent upon ALL the fifteen assumptions listed being warrantable.

Recently, two important papers on subcellular fractionation of Na-K ATPase have indicated the overall recovery of the enzyme in the crude particulate fraction as being only 20 per cent of that in the original homogenate (Kurokawa, Sakamoto & Kato, 1965; Rodnight, Weller & Goldfarb, 1970). This is clear evidence — as the authors point out — of considerable loss of enzyme activity during preparation. Many authors have obscured this loss by expressing the activities of fractions as a percentage of the total activities added together, which have survived homogenization.

Control experiments for the concentrations of enzymes in subcellular fractions

Many such experiments have been mentioned in the description of the steps of subcellular fractionation. The only common control experiments at present carried out are the overall recoveries of enzyme activity in subcellular fractions, compared with the enzyme activity in the crude homogenate. Nevertheless, there are several controls which are possible for the whole of the process. These are:

(i) the use of boiled tissues as parallel controls for heat-labile enzymes, and subtraction of the values of the apparent enzymic activity of these tissues from the substrate breakdown of the unboiled fractions; these would seem to be absolutely urgent and vital;

(ii) the division of subcellular fractions into two aliquots, one of which should again be subjected to the whole subcellular preparation system, the test of the change in activity of the latter, and the correction for any loss during the preparation;

(iii) the test of the effect of the preparative procedure on bacteria producing exo-enzymes; this would give an opportunity to measure the effect of subcellular fractionation quantitatively;

(iv) the mixture of boiled bacteria of known and uniform size with different enzymes in order to control such variables as enzyme and product extractability;

(v) the admixture of purified enzymes with synthetic polyamino-acids of known composition and size, to try to derive empirical relationships between the sizes and chemical properties of these well-characterized compounds, on the one hand, and the enzyme activity as a result of a particular fractionation procedures, on the other;

(vi) the subjection of the pure enzymes in dilute solution, but mixed with suspensions of inert *particles* of known size, to the whole preparative procedure, to test the effect of friction of particles of known size; and

(vii) the subjection of native proteins, like egg albumen and plasma proteins, to the whole procedure, and examining such properties as their viscosity, their absorption, their binding powers, their pH, and their immunological reactions before and after.

Which of these control experiments are relevant or necessary will depend mainly upon the 'enzyme' activities one is attempting to measure (see page 117).

Albertsson, P. A. (1962) *Arch. Biochem Biophys. Suppl.,* 1, 264.
Albertsson, P. A., & Nyns, Ed. J. (1961) *Arkiv Kemi,* 17, 197.
Dixon, M., & Webb, E. C. (1962) *The Enzymes,* London, Longmans, page 422.
Koch, B., & Lindall, A. W. (1966) *J. Neurochem.,* 13, 1231.
Kurokawa, M., Sakamoto, T., & Kato, M. (1965) *Biochem J.* 97, 833.
Lolley, R. (1963) *J. Neurochem.,* 10, 665.
Moore, W. V., & Lindall, A. W. (1970) *J. Neurochem ,* 17, 1665.
Morton, R. K. (1955) in *Methods in Enzymology,* Vol. 1, ed. by Colowick, S. P., & Kaplan, N. O., New York, Academic Press, page 25.
Oster, G., & Pollister, A. W. (1956) *Physical Techniques in Biological Research,* Vol 1, *Optical Techniques,* New York, Academic Press.
Rodnight, R., Weller, M., & Goldfarb, P. S. G. (1970) *J. Neurochem.,* 16, 1591.
Udenfriend, S. (1962) *Fluorescence Assay in Biology and Medicine,* New York, Academic Press.

2
Histochemistry

Histochemical techniques involve the following steps:

1. Killing the animal.
2. The dead animal cools.
3. The tissue is fixed or frozen.
4. It is dehydrated, freeze-dried or subjected to freeze substitution.
5. It is embedded.
6. The tissue block is cut.
7. The section is mounted.
8. The tissue is deparaffinized.
9. The section is incubated with substrate mixture.
10. The reaction is stopped.
11. The tissue is dehydrated.
12. The section is mounted.
13. The section is cleared.

In order to discuss general techniques in histochemistry, the steps used in Gomorri's (1952) method for the localization of acid phosphatase are used to give the order of the procedure. However, the discussion is not confined to this one technique.

The effect of killing the animal was dealt with on page 1 and the cooling of the animal on page 5.

3. Fixation or freezing the tissue

We may define the intention of fixation as the arrest of biochemical and structural change by the addition of a chemical agent or by freezing the tissue. The everyday experience that the histology of the tissue does not change for a long time after fixation, is its usual justification. There is a relative lack of information about the changes during fixation, for example. Usually it is recommended to proceed with tissue examination as soon after the sample has been obtained, to minimize 'autolytic' change. This precaution shortens the time between fixation and further chemical analysis, which would also obviate 'post fixation' chemical change.

However, although fixation may inhibit histological change, it cannot possibly stop diffusion of soluble enzymes, substrates, or co-factors. Therefore if any of these three groups of components diffuses from its original location, the apparent enzyme activity may be substantially altered. It might be asked reasonably if the reason why histochemical methods so often demonstrate enzymes in membranes, is that the enzymes, substrates or stains, adhere to the insoluble lipids and proteins, as the soluble components of the system have been washed away during the preparation.

The diffusion of insoluble compounds away from their original location has been tackled in two ways. One of them, pioneered by Behrens, is to use organic solvents into which only the lipids but not the water soluble constituents will diffuse. This approach has been mentioned in connection with subcellular fraction (page 10). The other approach, freezing tissues, has been used extensively. Freezing would prevent relocation of those soluble materials which crystallized — when the crystallization was complete. Until it was complete, the diffusion would be slowing down due to the cold but might be accelerated by the mechanical pressure of the shrinkage of the outer layers. If the tissue is frozen below the eutectic point of the sodium or potassium salt solutions (Katzman & Wilson, 1961), for example, then the constituents may be extracted, and the contents of those ions examined. Furthermore, as the salts crystallize out, the concentration of the other water soluble constituents increases and this would depress their freezing points even lower.

Even if one does cool tissue to the temperature of liquid nitrogen, at which it may be considered to be well and truly 'fixed', the addition of solvents to the frozen mass presents a further hazard. The heats of solution may be sufficient to dissolve out materials adjacent to those which it is intended to extract. This could be tested very easily, by examining the concentration in the solvents of materials other than those intended to be extracted, under conditions in which the solvents are used.

When tissue is fixed or frozen, little energy can be utilized by it. Thus, during the cooling, all substrates whose distribution *in vivo* depends upon energy supply — as the active transport of sodium ions or amino-acids does — will be rapidly relocated by diffusion. When one talks of 'rapid freezing', one is using a method which takes seconds or minutes (Dawson & Richter, 1950), because of the very poor heat conductivity of tissue (Spells, 1960), while cations, amino-acids, and acetylcholine, can diffuse within milliseconds (for review, see, for example, Curtis & Eccles, 1959; Eccles, 1964). When one measures the ionic gradients, for example, one 'fixes' the tissue with trichloracetic acid, *i.e.* one stops all metabolism, upon which the transmembrane gradient depends, and then claims to find out where it was before. This has been compared with taking the plug out of a bath and then trying to measure how much water there was in it before the plug was pulled (Hillman, 1966). All we can claim is that the more rapidly we do the fixation, the less the soluble constituents will be able to redistribute themselves. However, if the fixation or freezing takes seconds, and the redistribution takes milliseconds, we are in a region of uncertainty. We can, perhaps, never know the *status quo ante*.

Before we arrive at this conclusion, perhaps we might suggest it would be useful to try to *measure* rates of relocation of constituent materials within tissue. The other approach is to use metabolizing tissue in equilibrium only, for example, tissue slices after 'pre-incubation' — which are doing work against the incubation media. This will be discussed in greater detail (page 111). One simple approach (Sensenbrenner, Rendon, Hillman, Waksman & Mandel, 1968), was to use 'unfixed' tissue. In isolated ganglion cells, under microscopic vision, the above authors watched the development of colour within incubating cells due to aspartate amino-tranferase in a medium containing substrate and indicator; they observed the enzymic activity in the cytoplasm but not in the nucleus. They could not conclude that this was the real and only location of the enzyme, since such a conclusion would imply that the nuclear membrane was as permeable to substrate and stain, as was the cellular membrane. It also implied that other constituents within the nucleus would not have any effect on the reaction of the stain with the product of enzymic activity. The first two of these difficulties could be circumvented by injecting the substrate-stain mixture directly into the nucleus. Unfortunately this could not be done with spherical ganglion cells — trying to penetrate them with micropipettes is like chasing a golf ball with a walking stick. However, it can be done with isolated mammalian vestibular neurons (Cummins & Hyden, 1962), and therefore probably could also be done with fruit fly salivary cells, Mauthner giant cells, or amoebae, in which it would be reasonable to suppose that the cell's metabolism could survive the rude intervention of a micropipette. This seems to be an attractively reasonable experimental approach.

In respect of fixation, perhaps the most frighteningly unlikely belief is that the fixative stops all the biosynthetic and metabolic reactions upon which the enzymic activity one is examining depends, but does not alter the activity of the one that is being measured. This is tantamount to alleging that the fixative knows the enzyme favoured by the interest of the research worker. Cold acetone and formalin have been estimated to decrease phosphatase activities by 90 per cent and 75 per cent (Gahan, 1967; Lodin, Faltin and Muller, 1970). Bruce Casselman (1959) summarized literature on enzyme activities surviving freeze-drying or fixation with cold acetone, ethanol, 10 per cent formol, and 10 per cent formol saline. In different conditions the following percentages of each enzyme activity survived: acid phosphatases, 2-79; alkali phosphatases, 0-91; esterases, 0-82; sulphatases, 12-74; cytochrome oxidase, 56; and succinic oxidase, 37. Formalin is known to act by reacting with NH_2 groups.

The cryostat is a closed chamber in which tissue can be cut while both it and the knife are refrigerated to between $-30°C$ to $-10°C$. This minimizes diffusion and lowers the initial temperature of the tissue and knife. Thus any temperature rise would start with the tissue at a lower temperature so that more heat could be tolerated before the system reached such a temperature that enzymes would be irreversibly denatured. It may be stated without reserve that the principle of freezing a tissue to minimize change within it during the preparation is preferable under all circumstances to attempting to achieve the same object by the use of powerful chemical agents.

Perhaps, this section should finish by reiterating that *no chemical reagent known can stop the diffusion of soluble compound along their electrochemical gradients*, unless it precipitates them, or 'binds' them to non-mobile components.

4. Dehydration, freeze-drying and freeze substitution

Dehydration is usually carried out by using a succession of solutions of increasing alcohol concentrations. The tissue is dipped or rinsed in relatively large volumes of these solutions, often for hours at a time. It may be assumed that most of the water and alcohol soluble constituents are washed out, leaving one with a shell consisting of materials insoluble in water. This has been discussed in detail already.

The dehydration causes an overall shrinkage of tissue by 25 per cent as the water diffuses out. This will be discussed in the section on electromicroscopic preparation. Suffice it to say here that the shrinkage of the different tissue components will be quite unequal, because they each start off with different water concentrations. Thus the density of each chemical component of the tissue will change to different extents and this will give a false impression of depth of colour due to regional differences in enzyme activity.

The use of alcohol, methanol, butanol, acetone or ethylene glycol, implies that one believes them to have no significant effect on the activity of the enzyme being studied.

Freeze drying is, of course, carried out to fix tissue as well as to dehydrate it. Although they are dealt with here under the same heading, there is abundant evidence that the chemical and enzymic picture is not the same for chemically fixed and freeze-dried preparations (Bloom, Swigart, Scherer & Glick, 1954; Mayersbach, 1957; Yokohama & Stowell, 1951).

Theoretically, freeze drying or freeze substitution should permit the least movement of small soluble tissue components, and both of them preserve some enzyme activity, though the proportion of the original total is unknown.

5. Embedding

This is usually carried out in paraffin or in polyethylene glycols (Carbowaxes). The paraffin melts at 54-58°C, and Carbowax requires 47°C (Barka & Anderson, 1963, page 21). Bruce

Casselman (1959) quoted the percentage enzyme activities surviving from 20 minutes at 56°C to 2 hours at 60°C; these varied in the following way: acid phosphatases, 4-41; alkali phosphatases, 20-58; esterases, 10-42; cytochrome oxidase, 0, and succinic dehydrogenase, 0.

There is considerable literature on heat 'denaturation' of proteins, enzymes and nucleotides (see, for example, Joly, 1965), and these are the sort of temperatures which denature partially or completely. It is thus clearly necessary to see how these temperatures would affect the enzymes, as well as their histochemical detection.

Should one be suspicious of the fact that embedded tissues can be stored for years at room temperature without apparent loss of enzyme activity (Gomori, 1952; Chang & Berenbom, 1956)? Perhaps the absence of an aqueous phase preserves the enzyme. One should compare the storage properties at room temperature of the same enzyme mixed with tissue and dried. Would one not expect a purified enzyme to lose its activity at room temperature?

Barka and Anderson (1963, page 217) concluded that most authors agree that the greatest loss occurs during embedding.

6, 7. Cutting and mounting

Cutting wax – like boring Count Rumford's cannon – generates heat, and there will be a local rise in temperature. Whether the temperature rise is sufficient to affect the tissue cannot be determined easily (see page 15). It cannot be calculated without knowing the values of all the parameters quoted therein, and it would be difficult to measure as it is so localized and transient. We cannot even have an idea of its potential effect on a different preparation of the same enzyme because enzyme activities are measured after homogenization which is really a multiple chopping procedure. One could, however, take the reasonable precaution of cutting the tissue as slowly as possible, so that the heat generated will have enough time to be dissipated, before it induces a temperature rise.

Mounting the section consists of floating it on to a slide. This sometimes involves melting the paraffin and the same considerations apply to melting it at this stage as do during the embedding (see above).

8. Deparaffinization

Xylol or petrol ether can be used, and both of these will extract lipids. This would certainly be expected to alter the permeability of any biological membranes. In doing so, they might well render them permeable to enzymes, substrates and co-factors, to which their membranes had previously been impermeable. What other effects they would have on enzymes remains to be seen, and must be examined before any significant results can be claimed for these experiments.

Following treatment with the organic solvents, the tissue is usually washed with alcohol, and electrolyte, or distilled water. The usual apprehensions about these reagents removing soluble materials from the sections may be irrelevant at this stage, as all the latter materials may well have already been removed earlier on in the procedure.

9. Incubation

Tissue a few microns thick often has to be incubated for several hours. Some of the original enzyme activity will have been lost. What is supposed here is that the enzyme will have been lost in proportion to its original concentration at each of the tissues, *i.e.* that it was bound equally strongly to each subcellular component so the same proportion would be lost from each. This proposition is — to say the least — highly unlikely.

The substrate-buffer-indicator mixture would be of very different chemical composition than the original environment of the tissue. This difficulty will be discussed later (page 112).

Many histochemical incubations, like biochemical ones, are carried out under pH conditions far removed from those *in vivo*. This arises from the desire to detect 'optimum', *i.e.* maximum, enzyme activities. It may, however, give a false idea of their real locations. The theoretical question of the measurement of maximum enzyme activities will be dealt with in greater detail (page 111).

Quite often, incubation is carried out not in the natural substrate but in a substrate whose breakdown can be coupled to a dye, or one of the original substrate's relatives, or in the presence of an inhibitor of alleged 'specific' breakdown of the substrate (for review, see Augustinsson, 1950). For example, naphthyl acetate is used to detect cholinesterase, and the naphthyl liberated is coupled to a diazonium salt (Nachlas & Seligman, 1949). To use this technique, one has to believe that – not only in the standard conditions in which these authors demonstrated it – there is a parallelism between the cholinesterase activity and the breakdown of naphthyl acetates, but that this parallelism will be found in all the different kinds of preparations in which this technique may be used to detect the enzyme activity. Acetylthiocholine is used as a substrate of acetylcholinesterases, although it is hydrolyzed more rapidly than acetylcholine (Koelle & Friedenwald, 1949). 'Specific' inhibitors – of which a great number have been described for cholinesterases – are used to prevent the breakdown in tissues. These inhibitors have been shown to decrease the activity of enzymes in homogenized preparations, and they certainly increase the intensity of staining for enzymes. The meaning of 'specific' in this sense needs some clarification, but it is beyond the ambit of the present discussion (see page 118, (l)). Perhaps one can quote the quip attributed to Davenport, that 'the specificity of an inhibitor is in inverse proportion to the intensity of the investigations which have been done on it'.

10. Stopping the reactions

This usually involves lowering the temperature of the section to that of the room (18-25°C), and washing it with ice-cold fixative or water. All these measures will, of course, lower enzyme activities to a very slow but finite rate. Once again, it must be stressed that chemical agents can inhibit reactions, but cannot stop physical processes, like diffusion or osmosis.

In stopping the reactions, one is also washing off excess substrate mixture, so that what is left is the reagent which has been chemically, electrostatically or metabolically associated with the particular region of the cells. Although a histochemist believes that his substrate gravitates naturally to its home enzyme, all that can be asserted without fear of contradiction, is that following the several measures that have been taken, a substrate which may have been assumed to be normal for the enzyme examined has adhered to a particular subcellular organelle despite the multiple assaults of powerful reagents, all intended to diminish or relocate every other chemical which would otherwise react with it.

In view of the large number of reagents involved, it would probably be desirable, if possible, to incubate the tissue in media containing optimal electrolyte, glucose, buffer and oxygen concentrations much earlier on in the procedure, *i.e.* before 'fixation'. Incubated tissue has a very disorganized oedematous appearance histologically (H. Hillman, 1968, unpublished).

11, 12. Dehydration and mounting

The same considerations apply to these two stages as were mentioned under 4 and 5, except that the tissue may have been so deprived of its water, alcohol and xylol soluble materials in the earlier stages, that these further treatments can extract no more.

13. Clearing

This is usually done with clove oil or acid-alcohol, further agents whose effect on enzyme activity and localization would merit study. Suffice it to say that in altering the optical properties of a system, we must also be altering its physical chemistry.

Assumptions implied in use of histochemical techniques

(a) There is little post-mortem redistribution or loss of enzyme activity.*
(b) 'Fixatives' stop overall biochemical activity of tissue, but not the enzyme being studied.†
(c) The effects of freezing, *e.g.* shrinkage and intracellular crystallization, do not produce irreversible changes in tissue.
(d) Different parts of the tissue are dehydrated equally.*
(e) Warming tissue to embed it affects neither the localization nor the activity of the enzyme measured.†
(f) That cutting embedded tissue does not cause a significant temperature rise.

* this assumption contradicts laws of thermodynamics or physics.
† in some experimental systems, this has been shown to be untrue.

(g) That agents such as formalin, alcohol, or propylene glycol, do not affect the enzyme activity.†
(h) That incubation in aqueous media does not relocate the enzymes or change their apparent activity.*
(i) That 'non-specific' uptake of substrate, enzyme product, co-enzyme, stain or indicator does not occur.
(j) That the location or apparent concentration of enzyme is not affected by incubation in unphysiological media.
(k) That the staining reaction is not affected by the washing or clearing reagent.

Postmortem changes — assumption (a) — cannot be avoided, but they can be lessened by rapid handling of tissues, and by cooling them (Kerr, 1935; Stone 1938). Nevertheless, one is peeping into the twilight of uncertainty, in that death itself causes changes in tissue permeability to large molecules, which must then flow along their chemical gradients.

It would seem to be both absolutely necessary and relatively elementary to examine the enzyme activities and rates of diffusion both in relation to cold and all common fixatives used. By how much are the enzyme activities affected irreversibly by these two agents? Fixatives 'denature' proteins. Why do they not 'denature' the enzymes we are studying? Are enzymes exempt from cold 'denaturation'? What is the permanent effect of the extreme freezing used for rapid histological studies? If there is a significant effect, a mathematical correction may be made for it in our measurements.

Perhaps the most naïve assumption of all, (b), implies that the fixatives somehow know the reactions which the histochemist is studying. They thus arrange to inhibit all other enzymes almost completely, except for the one in which we are interested. Of course, it would not matter if some tissue enzymes were inhibited, but the annoying ones are those upon which the enzymes that we are studying are dependent for their rates. The worthiness of this assumption should have been doubted since it has been known that enzymes drive substrates along pathways, and ride cycles. The verbal justification given for this assumption, that histochemistry

---

* this assumption contradicts laws of thermodynamics or physics.
† in some experimental systems, this has been shown to be untrue.

is not expected to be quantitative, can hardly be taken seriously.

The effect of dehydration will be dealt with under the heading of electron microscopy (page 54). Examination of the effects of all other agents, such as alcohol, stains and clearing agents, used in a preparation for enzymic detection would again seem to be an elementary necessity. One would feel guilty about the constant insistence on it, if anything like a comprehensive study of their effects had ever even been undertaken.

The assumption, (h), that incubation in aqueous media will not relocate water soluble enzymes has already been mentioned as contrary to the Second Law of Thermodynamics. Assumption (j) is a different heresy, that 'non-specific' uptake will not occur according to the usual chemical laws governing diffusion, affinity, binding etc. There seems to be no reason why biological tissue with its cornucopia of covalence and conjugates, should be exempt from the usual laws on the chemical statute book governing the relations between atoms and molecules.

The essential evaluation of histochemistry depends upon two distinctions, both of which are examined elsewhere. Are we detecting an enzymic activity or the breakdown of a substrate (see page 117)? Are we interested in the properties derived from the life of the tissue, or are we evaluating the properties of organic material which has been grossly processed from dead tissue (see page 103)? Or are we looking at the properties of a changing condition in between?

The fact that these enzyme locations are detected 'optimally' by such complicated, particular and empirical techniques, might indicate that our measurement of their local concentrations may reflect the technique of preparation, rather than their localization *in vivo*. It seems entirely conceivable that a different incubating or staining technique might put the enzyme quite elsewhere. The observation has already been made that it is rare to detect enzymes by histochemical methods in the extracellular space or the cytoplasm.

One is continuously faced with the difficulty that chemical activities, especially of large organic molecules, are determined by nearly every other chemical species, with its own quantal idiosyncrasies, sitting around and affecting the lives of the species upon which biochemists happen so unpredictably to have turned their attention. Putting it a different way, the chemical activity of a molecule is a function of its total environment and it changes as that environment alters.

Of course histochemistry may give useful, if empirical, results when exactly the same technique is employed to examine the same tissue subjected to two different conditions. Even then, one would have to show, rather than just believe, that none of the stages of the procedure was substantially affected by the experimental situation.

One should add here that although all the strictures applicable to histochemistry are also theoretically relevant to morbid histology, in practice the control normal tissue has been subjected to precisely the same procedure. Furthermore, the interpretations, for example, used in the diagnosis of carcinoma in a rapidly cooled biopsy sample, are ones derived by observing empirically the correlations between the histological appearance, the malignancy, and the rate of spread of the disease. These correlations are happily independent of theoretical considerations.

Control Experiments

Many of the problems associated with histochemistry could be circumvented by appropriate controls, some of which are already being done. Experiments to examine the effect of each of the steps are mentioned under the respective headings. The following experiments would test the whole procedure:

(i) use of boiled tissue as controls especially for known labile enzymes;
(ii) use of tissues irradiated with ultraviolet light, if it inhibits that particular enzyme, as a control;
(iii) leaving out the substrate as a control;
(iv) incubation with inhibitors exhaustively shown to be 'specific';
(v) preparation of 'dead' organic materials, like wool or wood, in the same way as is done for histochemistry. This would not be expected to have enzymic activity but would be relevant to their localization by histochemical preparation.

Augustinsson, K-B. (1950) *The Enzymes*, Vol. 1, ed. by Sumner J. B., & Myrback, K., New York, Academic Press, page 443.
Barka, T., & Anderson, P. J. (1963) *Histochemistry*, New York, Hoeber Medical Division.

Bloom, D., Swigart, R. H., Scherer, W. F., & Glick, D. (1954) *J. Histochem. Cytochem.*, 2, 178.
Bruce Casselman, W. G. (1959) *Histochemical Techniques*, London, Methuen, page 164.
Chang, F. P., & Berenbom, M. (1956) *Exp. Cell. Res.*, 10, 228.
Cummins, J., & Hyden, H. (1962) *Biochem. Biophys. Acta*, 60, 271.
Curtis, D. R., & Eccles, J. C. (1959) *J. Physiol.*, 145, 529.
Dawson, R. M. C., & Richter, D. (1950) *Amer J. Physiol.*, 160, 203.
Eccles, J. C. (1964) *Physiology of Synapses*, Berlin, Springer, page 42.
Gahan, P. B. (1967) *Int. Rev. Cytol.*, 21, 1.
Gomorri, G. (1952) *Microscopic Histochemistry*, Chicago University Press.
Hillman, H. (1966) *Int. Rev. Cytol.*, 20, 135.
Joly, M. (1965) *A Physico-Chemical Approach to the Denaturation of Proteins*, New York, Academic Press.
Katzman, D., & Wilson, C. E. (1961) *J. Neurochem.*, 7, 113.
Kerr, S. E. (1935) *J. Biol. Chem.*, 110, 625.
Koelle, G. B., & Friedenwald, J. S. (1949) *Proc. Soc. Exp. Biol.*, 70, 617.
Lodin, Z., Faltin, J., & Muller, J. (1970) *Acta Histochem.*, 35, 43.
Mayersbach, H. (1957) *Acta Anat.*, 30, 487.
Nachlas, M. M., & Seligman, A. M. (1949) *J. Nat. Cancer Inst.*, 9, 415.
Sensenbrenner, M., Rendon, A., Hillman, H., & Waksman, A. (1968) *J. de Physiol. Paris*, 60, suppl. 1, 303.
Spells, K. E. (1960) *Physics in Biol. & Med.*, 5, 149.
Stone, W. E. (1938) *Biochem. J.*, 32, 1908.
Wolman, M. (1965) *Int. Rev. Cytol.*, 4, 79.
Yokohama, H. O., & Stowell, R. E. (1951) *J. Nat. Cancer Inst.*, 12, 211.

3

Electronmicroscopy

When a tissue is prepared for electronmicroscopy the steps followed are:

1. Killing the animal (page 1).
2. The dead animal cools (page 5).
3. The tissue is 'fixed' (page 41).
4. It is dehydrated (page 45).
5. The cytoplasm is replaced by a non-aqueous solvent.
6. The tissue is embedded (page 45).
7. It is sectioned (page 46).
8. It is stained (page 47).
9. It is mounted (page 46).
10. It is subjected to an electron beam.

Most of these steps have been examined on the pages indicated in brackets in the section on histochemistry, but a few are more particular to this technique.

3, 4. Fixing and dehydration

When tissue is fixed for electron microscopy, the appearance of the cells and their organelles depends upon the fixative, whether it be an aldehyde, osmic acid or permanganate. It also depends upon the pH, the fixative vehicle, the total ionic concentration, the dielectric constant, the osmolality and the specific ionic composition (for review, see Trump & Ericsson, 1965). In order to assess each of these factors, Trump and Ericsson examined the microstructure of rat kidney tubules, which have been well characterized by light microscopy. Studies on aldehydes, osmium salts and permanganate, with different buffers showed how the appearance of membrane, microbodies, mitochondria, endoplasmic reticula, cytoplasm, and nuclei, depended upon the particular conditions of fixation.

Osmium salts have been shown to extract about 15 per cent of the tissue nitrogen from kidney slices, and about 30 per cent from isolated liver mitochondria (Dallam, 1957). They react strongly with lipids (Adams, 1960) and unsaturated fatty acids (Stoeckenius and Mahr, 1965), and very little with tissue proteins (Bahr, 1954, Hake, 1965). It is thus clear that osmiophilic structures will show up much better than osmiophobic ones.

During fixation shrinkage occurs. A whole monkey brain will decrease in volume 11-25 per cent, whatever fixative is employed (Frontera, 1959). Single cells seem to fare much worse. With six different fixatives used in conventional histology, tissues shrink to different extents (Lodin, Mares, Karasek & Skrivanova, 1967), 25-35 per cent on average (Frontera, 1959). An attempt to decrease shrinkage in electron microscopy by freeze substitution was made by van Harreveld, Crowell and Malhotra (1965). These authors evidently believe that it is possible to prevent shrinkage during dehydration. Since living biological tissue is 60-70 per cent water, the organelles would each shrink *in the same proportion*, only if the membranes, the mitochondria, and the cytoplasm, originally contained the same concentration of water *in vivo* – an unlikely proposition.

During fixation and staining of tissues, a very thin film of fixative is deposited on the whole of the surface of the structure. If we imagine a very fine solid single-layered membrane, say a few hundred Ångstrom units thick, upon which we shower osmium salts in a fine rain, it would give a uniform thickness of deposit, like a rope on a tarry quay. When the original membrane was evaporated off (see page 59) our *single-layered* membrane seen on plan would appear to have *two* walls of salt descending into the plane of the section on either side of it, as if the rope had been removed; the edges on plan would appear as a *thicker* deposit than the osmium in which it was lying (Preuss, 1965, and see below, page 65). Thus any single thickness fine structure would appear double layered. Is there an alternative model possible which would give a single-layered image of a longitudinal section of a fine structure? The only one which comes to mind is if there were a solid structure adjacent to a space, in such a manner that only one wall would have the fixative or stain deposited on it.

One of the most popular uses of electronmicroscopy has been the measurement of intermembrane distances, and the sizes of 'subcellular' structures. Such measurements have been adduced in calculations of extracellular spaces, and rates of transport across membranes and junctions (see, for examples, Maynard, Schultz & Pease, 1957; Horstmann & Meves, 1959; Smith, 1963; van Harreveld, Crowell & Malhotra, 1965). The extracellular space in mammalian brain calculated from electron micrographs has usually been from 10-20 per cent less than that estimated chemically (Manery, 1953; Woodbury, Timiras, Koch & Ballard, 1956; Davson & Spaziani, 1959; Aprison, Lukenbill, & Segar, 1960). The discrepancy may be due to the fact that the latter estimates were carried out on living animals, while the former were on fixed dehydrated tissue sections. Although differential shrinkage makes absolute measurements of lengths and volumes from electron micrographs meaningless, interference or phase microscopy of cultured unfixed cells seems to be permissible (see, for example, Buchsbaum, 1948; Edwards, Paule & Pomerat, 1964; Lodin et al. 1967). However, cells *in vitro* may be of different dimensions than they are *in vivo*, because they are not subjected to the hydrostatic pressure due to a column of tissue, or the blood pressure.

4, 5. Dehydration and cytoplasm replacement by non-aqueous solvents

Tissue is dehydrated by the application of solvents which can be used to extract lipids. We must, therefore, believe that fixation of the tissue must protect most of it from these organic solvents. This may be due to the formation of organo-metal complexes, which are not so soluble in organic solvents; or it may be due to incomplete extraction of the lipids by the solvents. It might also be appropriate to remind ourselves at this stage, that during the initial aqueous steps of the dehydration, unknown proportions of the water soluble proteins, hormones, amino-acids, fatty acids, carboxylic acids, and vitamins, are lost. The implications of these facts is that electronmicroscopy — like subcellular fractionation and histochemistry — gives us information mainly about water and alcohol insoluble parts of cells (see page 10).

During dehydration, all the insoluble cytoplasmic constituents, which have not diffused away will precipitate out. *Any* constituents within the cytoplasm – crystals, particles, mitochondria, and ribonucleoproteins, for example – will deposit; indeed, all insoluble particles will remain like the stones carried onto the beach as the waves recede. These materials may precipitate in ordered rows, especially if they are charged. Crystals of L-glutamate dehydrogenase, and ox liver catalase, show beautiful layered structures on electronmicroscopy (Hall, 1950, 1960). Negative staining of catalase crystals has demonstrated, most elegantly, layers with a period of 100 Ångstrom units (Horne, 1965). It would seem legitimate to wonder whether the two kinds of 'endoplasmic reticulum' are something other than the flotsam and jetsam on the cytoplasmic shore.

The electron microscope must be reduced to a very low pressure indeed before the electron beam is directed on the tissue (see below). This, with the heat due to the electron beam will evaporate most of the organic material, as it can be seen that the embedding medium sublimes.

Measurement of the sizes of subcellular organelles depends upon determination of the magnification: this was first carried out by comparison of a 20 μ hole under light microscopy with the image under electronmicroscopy (von Borries & Ruska, 1939). Gratings (Burton, Barnes & Rochow, 1942) and crystal lattices (Menter, 1956) have also been used, and the total error with these techniques is about 5-10 per cent (Bahr & Zeitler, 1965). The materials used were glass or metal, which are relatively stable, but latex spheres appeared to be very unsatisfactory, due to contamination, swelling on standing, determinations being dependent on beam intensity (Watson & Grube, 1952), coalescence, and deformation (Bahr & Zeitler, 1965). This indicates that it would be extremely difficult to measure the size of subcellular particles on electron microscopy.

The calibration mark on an electron micrograph indicates the measured size of the marker *before* it was put into the microscope. If the shrinkage is considerable – as it is with tissue, but not with the inorganic magnification standards – the calibration becomes irrelevant. Unfortunately, one cannot even use it to measure *relative* sizes, since the different cellular components must shrink to different extents (see page 56).

It is often said that electronmicrograph measurements on dehydrated tissue are in agreement with the low angle x-ray diffraction analyses of Finean (1958), but, alas, these analyses, also, are carried out on partly dehydrated tissues.

One way of calibrating the electronmicrograph for the measurement of the sizes of cells, nuclei and mitochrondria, would be to measure unfixed bacteria of similar sizes and shapes to each of these organelles by light microscopy (see Hyden, 1961) and then measure them again after preparation for electronmicroscopy; this would give a real correction factor.

7, 8. Sectioning and staining

The tissue is sectioned, and this generates heat (see page 46). Distortion of the shapes of some particles has been studied (Peachey 1958; Satir & Peachey, 1958) and equations for corrections have been derived (Loud, Barany & Pack, 1965). These distortions can sometimes be recognized on the micrographs.

The staining of tissues for electron microscopy requires agents, which attach to particular compounds. The affinity of osmic salts for lipids has been mentioned in connection with fixation. DNA can be shown up with indium trichloride or uranyl acetate (Watson & Aldridge, 1961; Huxley & Zubay, 1961). Basic amino-acids react with phosphotungstic acid (Kuhn, Grassman & Hofmann, 1957; Hodge & Schmitt, 1960). One must believe that the relative specificity of these agents for compounds is the same in simple mixtures, as it is in the tissue section as prepared for staining (Beer, 1965).

10. The electron beam

The stained tissue is placed on a grid and the electron microscope evacuated down to pressure of 10^{-5} to 10^{-4} torr. It is then subjected to an intense beam of electrons which produces radiation damage, heat change, and evaporation. Reimer (1965). Kobayashi and Sakaoku (1965), Bahr, Johnson and Zeitler (1965), and Reimer and Christenhusz (1965), produced four extremely useful summaries of these effects from which the following generalizations have been culled.

Temperatures of the specimens often rise 500 to 1000°C (this is the range which is often used to ash tissue, before estimating the quantity of inorganic salts, for example). The vacuum in the tube will aid evaporation and prevent heat escape except by radiation.

After irradiation, organic polymers used for embedding, like araldite, methacrylate, or polystyrene, can lose 20 per cent, 50 per cent or 5 per cent, of their masses, and 30, 33, or 18 per cent of their carbon contents, respectively; these are just three examples from the Table 2 drawn up by Reimer (1965). Hydrocarbons, alcohols, ethers, aldehydes, ketones, esters, carboxylic acids and halides, may be degraded. Optical and ultraviolet absorption in certain peaks of the spectra of organic compounds may decrease up to 80 per cent and, of course, by no means uniformly along the spectrum. Kobayashi and Sakaoku (1965) illustrate photographs of the electron diffraction patterns of polyethylene crystals as a function of the temperature from 150 to 1000°, of exposure up to three minutes, and of voltage between 75 and 300 kV.

Some further considerations in respect of electronmicroscopy

It would be reassuring to believe that despite all the dangers of electronmicroscopy previously listed, there was a consensus of consistent information which had been learnt by the use of this technique. Let us examine some currently held beliefs.

(a) The existence of endoplasmic and sarcoplasmic reticula.

Two kinds of structures have been detected in the cytoplasm: a 'rough' and a 'smooth' reticulum (Porter, Claude & Fullam, 1945; Porter, 1953; Palade & Porter, 1954; Palay & Palade, 1955). Observed early in liver and neural tissue, the reticulum has been seen in nearly all types of animal and plant cell (see, for example, Hurry, 1965; Fawcett, 1966). The structure is now so universally accepted that Hurry's is a textbook issued for senior schoolchildren.

The electronmicrographs published show a distance between strands of the three dimensional reticulum of 800 to 2000 Ångstrom units (Å), while the nuclei are 50,000 to 200,000 Å, and the smallest dimensions of the mitochondria are between 5,000 and 20,000 Å. All observers of tissue cultures see the nuclei, mitochondria, and other light microscopically visible particles, continuously moving about in the living cells (see, for example,

Costero & Pomerat, 1961; Hansson & Sourander, 1964; Willmer, 1965). How is it possible for these structures to swim around in a three dimensional net, one that is three orders smaller than they are? Furthermore, this honeycomb — as I would prefer to call it — is often accepted as attached both to the nuclear and the outer cell membrane (Watson, 1955; Palade & Siekevitz, 1956); this would also make movement of organelles quite impossible. Streaming can also be seen in adult cells *in vivo*, when they can be viewed suitably.

The circumstances in which the endoplasmic reticulum could exist, while the organelles moved, would be either if the endoplasmic reticulum did not permeate the whole cytoplasm, or if it stretched as the organelles moved. In the latter case, it would be reasonable to expect to see the leading edge of the moving organelle compressing the reticulum and the trailing edge stretching it. That would give a quite characteristic appearance, which I am not aware has been seen; the closest to it that is frequently observed are the compressed layers of the reticulum, which is the Golgi body. That could be in front of the organelle, but what would be behind it? Has anyone reported elongated strands in the direction of the movement but trailing the organelle?

We cannot ignore the evidence that the endoplasmic reticulum on electron micrographs looks similar to the electron micrographs of the cellular fraction of the same name. Their similarity in histology is extremely difficult to assert in view of their unstructured appearance; their apparent similarity in size is a visual impression rather than an experimentally measured, statistically valid, comparison. There is a considerable literature on their histochemistry and subcellular chemistry (see, for example, Palade & Siekevitz, 1956; Reid, 1967; Tata, 1969); this has yet to be evaluated critically.

If the endoplasmic reticulum is indeed an organized structure *in vivo*, what happens to the cytoplasmic constituents after dehydration? Do they diffuse away, evaporate, deposit or disappear? The appearance of endoplasmic reticulum seems most likely to be due to precipitated cytoplasmic solutes. Otherwise, where else do they go?

If the endoplasmic reticulum is attached to the extranuclear space, and, or, the extracellular space (Palade & Porter, 1954; Palade, 1955), and if the cell membrane is two layered, why is the reticulum not four-layered? To where have the other two layers disappeared? If it is attached to the extranuclear space, the nucleus, which can be seen to rotate in any tissue culture, must be enclosed by a separate one- or two-layered structure; thus, in those regions of the nuclear surface where the alleged pores are not joined to the endoplasmic reticulum, one should expect to see a three- or, possibly, four-layered nuclear membrane. It has also been said that the endoplasmic reticulum is attached to the extracellular space and the nuclear space (Callan, 1950; Watson, 1955), and the endoplasmic reticulum is supposed to have a channel 200-500 Angstroms wide (Callan, 1950; Palade, 1955). Why then do not all the small molecules interchange between the extracellular fluid and the nucleus? Do we believe that the *milieu interieur* bathes the nucleus as well as the whole cell? If this were so, the resting potential would be the same in the nucleus as the cytoplasm; in Xenopus oocytes, it is, indeed, zero, while in Drosophila salivary gland it is 13 mV more; the transmembrane conductances are also different, yet the pores are apparently the same size on electronmicroscopy (Lowenstein & Kanno, 1963a, b; Kanno & Lowenstein, 1963; Wiener, Spiro, & Lowenstein, 1965; Lowenstein, Kanne & Ho, 1966). If to the endoplasmic reticulum connecting the extracellular fluid and the nucleus we add the pores between the nucleus and cytoplasm (see below), we have the three compartments in communication. Then, when we penetrate the cell membrane from the extracellular fluid, there should be no potential difference – but there is – and when we further insert the electrode into the nucleus, there should also be no potential difference – but there is. Furthermore, with such large and numerous holes as have been postulated, the nucleus and cytoplasm would have the same concentration of all diffusible products. What function can these membranes with all their holes have?

(b) The pores in the nuclear membrane.

Two kinds of pores have been described: the ones in the outer membrane through which small ions are said to diffuse (Solomon, 1960), and the ones in the nuclear membrane (Lanzavecchia, 1962), with which we are concerned here. Holes have been reported in many nuclear membranes of which the following are a few examples; their diameters (A) are indicated: rat neurons, 280-360 (Palay & Palade, 1955); rat spinal ganglion cells, 700 (Hartmann, 1953); various rat cells, 300-500 (Watson, 1955); mouse pancreatic acinar cells, 200-400 (Watson, 1954); amphibian oocytes, 500 (Callan, 1950; Feldherr, 1954); sea urchin oocytes, 1000 (Afzelius, 1954; Robertson, 1960);

The chlorophyll molecule has a diameter of about 59 A, gold particles, 125-150, porphyrins, 30-36, serum albumin, 70, globular proteins, 40, and simple proteins, 7-18. Organic molecules of molecular weights less than 100,000 have diameters less than 35 (Sobotka, 1944, and sources quoted there). Harris (1960) lists 18 ions of biological significance, including potassium, sodium, chloride, hydrogen, calcium and magnesium; he gives the diameters of their hydrated species, some of the values being from Conway, (1952); *they are all between 3 and 9 A*. How is it possible that such small ions do not diffuse through the pores (Harding & Feldherr, 1959; Lowenstein, Kanno & Ho, 1966) or even the endoplasmic reticulum channels, which are said to have a diameter of 100-200 A (Robertson, 1960)? No doubt the concept that the pores are charged would be adduced. Since the charge would have to decrease inversely as the square of the distance from the pores' edges, the metabolic pumps would have to do an enormous amount of work to keep the nucleus, the cytoplasm and the extracellular fluid different. There is plenty of evidence that the nucleoplasm is chemically different than the cytoplasm. It appears to concentrate sodium ions to some extent (Allfrey, Meudt, Hopkins & Mirsky, 1961; Langendorff, Siebert, Lorens, Hannover & Beyer, 1961).

The thickness of the section, the proximity of its edge, and the general state of the tissues, all determine the appearance of the organelles (Kay, 1967), besides the other factors already discussed. Furthermore, electronmicroscopists are well aware of the theoretical and statistical difficulties of comparing a small and variable aliquot of such a relatively large piece of tissue (Burge & Draper, 1965; Stoeber, 1965). Nevertheless, provided sufficient care is taken of the statistics and the two tissues are prepared in exactly the same way, significant differences between electronmicrographs can be found, especially in respect of properties which can be expressed digitally, like numbers of synapses (Cragg, 1967).

Electronmicroscopy

Assumptions implied by use of electronmicroscopic techniques

(a) Assumptions (a) – (f) as mentioned in connection with histochemical techniques (see page 49).
(b) That there are no significant structures which dissolve or diffuse away in the reagents.
(c) That there are no significant structures which are not stained.
(d) That the organelles originally are equally hydrated *in vivo*, and will shrink proportionately.†
(e) That the heat and irradiation causes no significant change in the size or shape of the organelles.†
(f) That the similarity of electronmicrographs from subcellular fractions, fresh tissue, and fixed tissue, means that electronmicroscopy has no effect on the preparation.
(g) That ability to distinguish organelles clearly on electronmicrographs is evidence of their biochemical viability before fixation.

Most of these assumptions have been dealt with in the text. Assumption (b), that structures might not dissolve or diffuse away in the reagents refers to all the aqueous solutions, as well as the dehydrating ones. The possibility of insoluble small particles or large molecules, which are loosely adherent to the 'larger' organelles just being washed away down the streams of reagents, does not seem to have been considered in the literature; such particles could be loosened from their moorings by glutaraldehyde, osmic acid, water, alcohol and propylene oxide – that is, by any of the reagents in which the tissue had come into contact before it is embedded.

Studies have been cited on the affinity of fixatives and stains for different components of the tissue. There does seem to be a good case for a systematic examination of organometal complexes with lipids, carbohydrates, and proteins, to see if agents other than osmic acid or phosphotungstic acid might be attached to other components of the tissue; the resultant complexes should be specified to be able to survive the assault of the electron beam, as

---

† These assumptions have been found experimentally to be untrue.

well as the embedding and vacuum. It seems to me entirely conceivable that a new range of fixatives and stains could be utilized for the whole spectrum of organic molecules present in biological tissue. The only reservation we must bear in mind is that a compound which reacts in a particular way with a pure solution, say of albumen, will not necessarily react the same way to the same chemical in its normal biological environment. Nevertheless, it might be possible to reveal a galaxy of hitherto undetected organelles.

Measurements of size made on electronmicrographs imply either that shrinkage does not occur, or that it is uniform throughout the tissue. The fact that the nucleus, the mitochondria and the membrane are not in solution in the cytoplasm means that they could not have the same water content as it. In the absence of knowledge of the relative extent of differential shrinkage in each organelle, one cannot know relative sizes by electronmicroscopy. Unless we are prepared to study the water content of the organelles on fractions isolated by homogenization and centrifugation — and pretend that we believe that such preparation would not alter the organelle volumes themselves — it may be that electronmicroscopy cannot be used for measurements of organelles.

One of the commonest assumptions (g), that a clear electronmicrographic image is evidence of tissue viability, could not really be farther from likelihood. It is *extremely* unlikely that the whole process of preparation from electronmicroscopy would not change any enzyme activity enormously, yet this criterion is used to justify the health of subcellular fractions in which enzymes and respiration are studied. It would be highly desirable to study the effect of an electron beam not merely on pure crystals of succinic dehydrogenase or cytochrome oxidase, but also on solutions of these enzymes.

Besides the control experiments mentioned in the text, the following experiments for overall examination of the electronmicroscopic technique would be highly desirable:

(i) the effect of the whole preparation procedure on materials of biological origin, like wool, leather, wood, and pollen, to identify the artefacts;
(ii) examination by light microscopy of the effects of the agents on the relative volumes of tissue components;

(iii) identification of the artefacts found on preparation of pure solutions and suspensions of enzymes, proteins, carbohydrates and fats;

(iv) when histochemistry is done with electronmicroscopy, boiled tissue or tissue with inhibitors should be used as controls;

(v) wherever possible, normal tissue and tissue from an animal subjected to a particular agent should be studied together. Most of the problems of electronmicroscopy can be avoided using this approach, but it may well be that the effects of preparation are sufficiently large in the experimental and control tissue to mask induced biological change; and

(vi) more general use of light microscopy of unfixed tissue (Hyden, 1961).

Adams, C. W. M. (1960) *J. Histochem. Cytochem.*, 8, 263.
Afzelius, B. A. (1954) *Exp. Cell. Res.*, 8, 147.
Allfrey, V. G., Meudt, R., Hopkins, J. W., & Mirsky, A. E. (1961) *Proc. Nat. Acad. Sci.*, 47, 907.
Aprison, M. H., Lukenbill, A., & Segar, W. E. (1960) *J. Neurochem.* 5, 150.
Bahr, G. F. (1954) *Exptl. Cell. Res.*, 7, 457.
Bahr, G. F., Johnson, F. B., & Zeitler, E. (1965) *Lab. Invest.*, 14, 377/1115.
Bahr, G. F., & Zeitler, E. (1965) *Lab. Invest.*, 14, 142/880.
Beer, M. (1965) *Lab. Invest.*, 14, 283/1020.
Buchsbaum, R. (1948) *Anat. Rec.*, 102, 19.
Burge, R. E., & Draper, J. C. (1965) *Lab. Invest.*, 14, 240/978.
Burton, C. J., Barnes, R. B., & Rochow, I. G. (1942) *Ind. Eng. Chem. Ind. Ed.*, 34, 1429.
Callan, H. (1950) *Proc. Roy. Soc. B.*, 137, 367.
Conway, B. E. (1952) *Electrochemical Data*, Amsterdam, Elsevier, page 62.
Costero, I., & Pomerat, C. M. (1951) *Amer. J. Anat.*, 89, 405.
Cragg, B. G. (1967) *Nature*, 215, 251.
Dallam, R. D. (1957) *J. Histochem. Cytochem.*, 5, 178.
Davson, H., & Spaziani, E. (1959) *J. Physiol.*, 149, 135.
Edwards, S. R., Paule, W. J., & Pomerat, G. M. (1964) *Anat. Rec.*, 148, 279.
Fawcett, D. (1966) *The Cell*, London, Longmans.
Feldherr, C. M. (1964) *J. Cell. Biol.*, 20, 188.
Finean, J. B. (1958) *Exper. Cell. Res.*, Suppl. 5, 18.
Frontera, J. G. (1959) *Anat. Rec.*, 135, 83.
Hake, T. (1965) *Lab. Invest.*, 14, 470/1208.
Hall, C. E. (1950) *J. Biol. Chem.*, 185, 749.
Hall, C. E. (1960) *J. Biophys. Biochem. Cytol.*, 7, 613.
Hansson, H. A., & Sourander, P. (1964) *Z. Zellforsch.*, 62, 26.
Harding, C. V., & Feldherr, C. M. (1959) *J. Gen. Physiol.*, 42, 155.
Harris, E. J. (1960) *Transport & Accumulation in Biological Systems*, 2nd edn. London, Butterworth, page 3.
Hartmann, J. (1953) *J. Comp. Neurol.*, 99, 201.
Hodge, A. J., & Schmitt, F. O. (1960) *Proc. Natl. Acad. Sci. U.S.*, 46, 186.
Horne, R. W. (1965) *Lab. Invest.*, 14, 316/1054.
Horstmann, E. M., & Meves, H. (1959) *Z. Zellforsch.*, 49, 569.

Hurry, S. W. (1965) *Microstructure of Cells*, London, John Murray.
Huxley, H. E., & Zubay, G. (1961) *J. Biophys. Biochem. Cytol.*, 11, 273.
Hydeh, H. (1961) in *The Cell*, ed. Brachet, J., & Mirsky, A., Vol. IV, New York, Academic Press, page 215.
Kanno, Y., & Lowenstein, W. R. (1963) *Exp. Cell. Res.*, 31, 149.
Kay, D. (1967) *Techniques for Electron Microscopy*, Blackwell, Oxford.
Kobayashi, K., & Sakaoku, K. (1965) *Lab. Invest.*, 14, 359/1097.
Kuhn, K., Grassman, W., & Hoffmann, V. (1957) *Naturwiss*, 44, 538.
Langendorff, H., Siebert, G., Lorenz, I., Hannover, R., & Beyer, R. (1961) *Biochem Z.*, 335, 273.
Lanzavecchia, G. (1962) *5th International Conference of Electron Microscopy*, Vol. 2, New York, Academic Press.
Lodin, Z., Mares, V., Karasek, J., & Skrivanova, A. (1967) *Acta histochem*, 28, 297.
Loud, A. V., Barany, W. C., & Pack, B. A. (1965) *Lab. Invest.*, 14, 258/996.
Lowenstein, W. R., & Kanno, Y. (1963a) *J. Gen. Physiol.*, 46, 1123.
Lowenstein, W. R., & Kanno, Y. (1963b) *J. Biophys. Biochem. Cytol.*, 16, 421.
Lowenstein, W. R., Kanno, Y., & Ho, S. (1966) *Ann. N.Y. Acad. Sci.*, 137, 708.
Manery, J. F. (1952) in *Biology of Mental Health and Disease*, New York, Hoeber, page 124.
Maynard, E. A., Schultz, R. L., Pease, D. C. (1957) *Amer. J. Anat.*, 100, 409.
Mentner, J. W. (1956) *Proc. Roy. Soc.*, A 236, 119.
Palade, G. E. (1955) *J. Biophys. Biochem. Cytol.*, 1, 567.
Palade, G. E., & Porter, K. R. (1954) *J. Exp. Med.*, 100, 641.
Palade, G. E., & Siekevitz, P. (1956) *J. Biophys. Biochem. Cytol.*, 2, 671.
Palay, S. L. & Palade, G. E. (1955) *J. Biophys. Biochem. Cytol.*, 1, 69.
Peachey, L. D. (1958) *J. Biophys. Biochem. Cytol.*, 4, 233.
Porter, K. R. (1953) *J. Exp. Med.*, 97, 727.
Porter, K. R., Claude, A., & Fullam, E. F. (1945) *J. Exp. Med.*, 81, 233.
Preuss, L. E. (1965) *Lab. Invest.*, 14, 181/919.
Reid, E. (1967) in *Enzyme Cytology*, ed. by Roodyn, D., New York, Academic Press, p. 321.
Reimer, L. (1965) *Lab. Invest.*, 14, 344/1082.
Reimer, L., & Christenhusz, R. (1965) *Lab. Invest.*, 14, 420/1158.
Robertson, J. D. (1960) *Prog. Biophys.*, 10, 343.
Satir, P. G., & Peachey, L. D. (1958) *J. Biophys. Biochem. Cytol.*, 4, 345.
Smith, K. R. (1963) *J. Comp. Neurol.*, 121, 459.
Sobotka, H. (1944) in *Medical Physics*, Vol. 1, ed. by Glasser, O., New York, Year Book Publishers, page 777.
Solomon, A. L. (1960) *Sci. American*, 146.
Stoeber, W. (1965) *Lab. Invest.*, 14, 154/892.
Stoeckenius, W., & Mahr, S. C. (1965) *Lab. Invest.*, 14, No. 6, 459/1196.
Tata, J. R. (1969) in *Subcellular Components*, ed. by Birnie, G. D., & Fox, S. M., London, Butterworth, page 83.
Trump, B. F., & Ericksson, J. L. E. (1965) *Lab. Invest.*, 14, No. 6, 507/1245.
Van Harreveld, A., Crowell, J., & Malhotra, S. K. (1965) *J. Cell. Biol.*, 25, 117.
Von Borries, B., & Ruska, H. (1939) *Naturwiss*, 27, 577.
Watson, M. L. (1954) *Biochem. Biophys. Acta.*, 15, 475.
Watson, M. L. (1955) *J. Biophys. Biochem. Cytol.*, 1, 257.
Watson, M. L., & Aldridge, W. G. (1961) *J. Biophys. Biochem.*, 11, 257.
Watson, J. H. L., & Grube, W. L. (1952) *J. Appl. Phys.*, 23, 793.
Wiener, J., Spiro, D., & Lowenstein, W. R. (1965) *J. Biophys. Biochem. Cytol.*, 27, 107.
Willmer, E. N. (1965) *Tissue Culture*, London, Methuen, page 7.
Woodbury, D. M., Timiras, P. S., Koch, A., & Ballard, A. (1956) *Fed. Proc.*, 15, 501.

4
Radioactive measurements

When radioactive measurements are made of rates or metabolic 'significance' of reactions in biological systems, the following steps are taken:

1. Injection or inhalation or addition of the radioactive isotope.
2. Uptake of the isotope by the tissue.
3. Its involvement in a metabolic pathway.
3. The enzymic activity is stopped.
5. The excess radioactivity is washed off.
6. The substances which have incorporated isotope are extracted, or subjected to autoradiography.
7. The compound structure, homogenate or extract containing the isotope is counted.
8. The results of counting are interpreted.

1. Injection of the isotope

Injection of a radioactive substance into an animal dilutes it tens to thousands of times. Thus the injected material must be very concentrated at the site of injection. Early on in the use of isotopes the biological effects of the radiation received on the body were recognized (for review, see Hevesy, 1948). The dangers are of two different types. One of them is the health hazard to the user, with which we are not concerned here (see Voigt, 1950), and the other is the possibility that the radioactivity itself would alter the properties of the tissue being studied. In an early study, Mullins (1939) showed that the permeability of Nitella was not affected by up to 1 μCi/1; Skanse (1948) found no effect for 48 hours on thyroid function of chickens given 50 μCi of radioactive iodine. In a more recent study, Puck (1960) irradiated cultures of S3 HeLa cells, Chinese hamster fibroblasts, and normal chick cells.

They were hardly affected by doses of 100 to 200 rads, and less than 1 per cent of cells were killed by 300 to 500 rads. Thus for most experiments there is little risk of radiation affecting the whole animal or an incubating tissue. There may be real risks both to the experimenter and to the clinical state of the animal if sufficiently active sample of a radioactive material is injected into a whole animal for single cells to be isolated for isotopic measurements.

2. Uptake of isotope by the tissue

The radioactive isotope will be distributed within the tissue according to its relative affinities to each of the constituents. The first experiments on radioactive uptake by tissues examined the distribution of polonium in rabbit (Laccasagne & Lattes, 1924). Later on, Hevesy and Hofer (1934) administered heavy water and Chiewitz and Hevesy (1935) used phosphate which had been made artificially radioactive. Whereas brain, medulla, spleen, kidney, and liver, took up (per ml of tissue) 14-18 per cent of the original specific activity, muscle took up 7.8, skeleton, 2.8, and blood only 1.8 per cent. Such studies are of value in measuring the uptake by tissue (see page 116), but tell one nothing about the rate, mechanism or kinetics of the systems involved. Nevertheless, even the empirical measurement of uptake depends upon the minimum detectable dose, the time since application of the material, the delay in removal of the sample, the uptake and excretion of the material, and the radiation characteristics and half life of the radiotracer.

At a tissue level, the uptake will depend upon:

(a) the exact site of injection;
(b) the total concentration of the isotope;
(c) the clinical state of the animal or metabolic state of the tissue;
(d) the blood flow *in vivo* or the concentration of the isotope *in vitro*;
(e) the length of time of exposure to the isotope;
(f) the specific chemistry of the isotope;
(g) the specific chemistry of the tissue;

(h) the pH at the site of uptake;
(i) the local temperature at the site of uptake;
(j) the reversibility of the chemical reaction between the isotope and the tissue;
(k) the rate of involvement of the isotope in metabolic pathways of the tissue by uptake or biosynthesis; and
(l) the rate of excretion of the isotope or the compounds with which it becomes metabolically related.

It should be noted that of this list, biosynthesis is only one of many factors. If we define 'specific uptake' as uptake or incorporation into a pathway in which we are interested, we have to show that this is the major route for this compound.

If one were to dip one's leg into a pool of radioactive water, one's foot, nails, shoe, sock and trouser leg would, no doubt, each take up different amounts of radioactivity. Their uptake would probably each occur at different rates. However, this would not mean that different components biosynthesized any product in which radioactivity persisted after washing, at these different rates; nor could one measure their 'relative specific activities' and 'turnover times'.

3. Involvement in a metabolic pathway

The isotope is involved in metabolic pathways. It joins a 'pool' (Schoenheimer, 1946; Sprinson & Rittenberg, 1949). The 'pool' is calculated by extrapolating the slope of incorporation back to zero time. It is a measure of the dilution of the isotope and is presumed to be related to the total amount of the non-radioactive isotope available for metabolism (Reiner, 1953; Arnstein & Grant, 1957). The pool may be constant; it may change during the killing of the animal done to carry out the measurement. It is at any moment a measure of the rate of mixing of the radioactive with the non-radioactive native isotope; it involves the balance of the flow *into* the system by diffusion, liberation and synthesis, and *out of* the system by diffusion, metabolism, degradation, and excretion (Francis, Mulligan & Wormall, 1959). Any agent could affect the pool size or the rate of incorporation by affecting any of these processes.

An excellent treatment of the mathematics of uptake of tracers is given by Bray and White (1966).

We can allege metabolism — as distinct from uptake — of a compound *in vitro*, if we do several control experiments; we must show that uptake of precursor does not occur in boiled tissue; that it is depressed considerably in the absence of substrate; that known inhibitors of those pathways — if such exist (see page 118) — or the absence of co-factors, decrease it considerably. Although most — if not all — of these control experiments would be necessary for unequivocal demonstration that the process was metabolically dependent, they are rarely done.

The difficulty is magnified *in vivo*, because these control experiments would so upset a living animal, that the massive change in its clinical condition could well be much greater than the difference expected between experiment and control. Furthermore, in the living animal, there are blood-brain barriers, detoxicating mechanisms, homeostatic mechanisms and regulatory systems, all resisting the violent changes which might be induced by any inhibitory agents — however certainly their action has been demonstrated on enzyme preparations. Thus, we can often assert honestly that control experiments cannot be done, (see page 77). Do these experiments have any certainty?

In many experiments using radioactive isotopes, the counts are compared in two different conditions. With the problem of chemical affinity for the isotopes (non-specific uptake), does a comparison of counts mean anything? One can get a very high activity in the tissue merely by using an isotope of high specific activity. The use of inhibitors which decrease the counts by say, 50 per cent, does not necessarily mean anything more than a change in location of uptake of the radioactive compound which could arise from a number of causes besides metabolism of the compound. Without knowing the 'non-specific' distribution of labelled compounds, as well as all the other pathways they could take, it is difficult to interpret a percentage change in counts. This may not always be apparent when a small dose of inhibitor, say a micromolar concentration, reduces an uptake by 30 per cent. The inhibitor concentration often appears to be comfortably small compared with the somewhat 'large' percentage change in the isotope uptake.

4, 5, 6. Enzymic activity stops. Excess radioactivity is removed. Incorporated substances are extracted

The stopping of the biosynthetic activity produces the same consequences as have been discussed under histochemistry (see page 48). Washing off the 'excess' radioactive precursor is intended to allow solution which has adhered to the outside of the tissue, or which has not been incorporated, to be removed without affecting that which has been involved in the metabolic pathway. Often a substance like trichloracetic acid is used to stop the reaction as well as to wash off the radioactivity. The assumptions behind this are, firstly, that it will stop the reaction at the rate it was at the moment when it was added, and secondly, that it itself will not extract or degrade significant quantities of the compounds in which the precursor has been incorporated.

Sometimes in the examination of tissue content of radioactive cations, we rinse the tissues 'rapidly'; this is intended to be sufficiently quickly so that the intracellular ions do not diffuse out and are not drawn out by surface tension. At the same time, it is meant to be sufficiently thoroughly so that nearly all the extracellular radioactive solution is diluted away. In other words, though we may recognize verbally that the mobility of the ions is not significantly different than in free solution – and the deprivation of the tissue's energy source would deny it any electrochemical mechanism for limiting this – we must beware of supposing that the ions inside can know that they are not meant to diffuse out. Thus, we must always have a very patient look at the effects of 'washing' and also try to find out how much 'contamination' there is. This term is used in the sense of how much radioactivity the extractant would pick up from an originally non-radioactive component, which was located physically or chemically near the component normally taken out by the extractant. This could be tested by counting the radioactivity of an extractant of the compounds other than the compound in whose fate one was interested. Perhaps also, the recovery of the radioactivity measured as a proportion of the total radioactivity in the body or tissue extracted – as well as that of the dose injected – would give a clearer indication of the significance of the radioactive material in the metabolic pathway.

The extraction of substances from mixtures has been dealt with under the section on subcellular fractionation (see page 31).

7. Counting

Counting radioactive isotopes is a complicated technology about which much has been written. Direct counting, scintillation counting, and autoradiography are the three main techniques, and each of these has been extensively dealt with (Pelc, 1951; Sacks, 1953; Francis, Mulligan & Wormall, 1959; Broda, 1960; Broda & Schonfeld, 1966). The count recorded will depend on:

- (a) the radioactivity of the source;
- (b) quenching by the tissue, except in liquid scintillation counting;
- (c) the radioactivity of the background;
- (d) the shape of the sample;
- (e) self-absorption by the sample;
- (f) refraction of light rays;
- (g) the geometry of the counter, except in liquid scintillation counting; and
- (h) the intrinsic efficiency of the counter.

The radioactivity of the source used will depend upon the isotope chosen, the size of the animal or tissue, the hazard to the experimenter, and, to a lesser extent, the ability to dispose of the radioactive waste. Its possible effect on the tissue with which it is in contact has already been mentioned. Quenching by the tissue is a rather complex property depending upon the tissue's chemical nature and physical state. It complicates the counting very much and ignorance about the magnitude of the parameters involved in a biological tissue inhibits any serious calculations. The same is true with respect to the shape of the sample, self-absorption, refraction of rays, and the geometry of the counter, in direct counting and in autoradiography (Pelc, 1951); these factors are of much less importance in liquid scintillation counting.

Difficulties are encountered with the calibration in radioactive counting experiments. Both 'external' and 'internal' standards are used. Clearly, allowance cannot be made for the geometry of the system when an external standard is used.

Two sorts of controls are possible. The usual and reasonable one is to compare tissues extracted and treated under exactly the same conditions, the one without the added agent under study being a control; in this case an external standard may be used as a check of the counter. The other, much rarer method, is to add a known activity to tissue samples of the same nature and plot the expected against the measured rate. This gives a real 'recovery' rate. It is, nevertheless, necessary to express extreme disquiet with the assumption that the radioactivity measurements will not be affected by the geometry of the tissue, as is implied by the use of external standards alone. In these connections, we must bear in mind that in counting, the efficiency cannot be assumed to be constant from day to day. Frequent use of a standard can indicate if the counter is relatively stable. It is, however, no substitute for the necessary 'recovery' experiments.

The efficiency of counting involves a further series of problems, which can be tackled in normal tracer techniques by the use of the external standard. Considerable efforts have been made in autoradiography to estimate the number of counts per grain; these have varied from 0.5 to 40 for a β emitter (for review, see Schultze, 1969, a, b). The properties of photographic emulsions are of great importance in autoradiography.

The 'purity' of radioactive precursors is an extremely important consideration. In 1948, Cohn summarized the sources of impurity and distinguished clearly between the chemical and radiochemical purity of a nuclide. Mullins, Fenn, Noonan, and Haege (1941) pointed out that a discrepancy between a 50 per cent exchange reaction after KCl bombardment with deuterons, and a 3 per cent exchange which Hahn, Hevesy and Rebbe (1939), had found, could be explained by a 0.02 per cent contamination with sodium chloride. Nowadays, radioactive isotopes are being prepared of greater and greater radiochemical purity as well as chemical purity, and this problem is no longer serious (Wilson, 1966).

Assumptions inherent in radioactive isotope rate measurement techniques

(Many of these can be found in Zilversmit, Entenman & Fishler (1943), and Siri (1949). Several of these assumptions are, however, not necessarily relevant to the *identification* of metabolic pathways.)

(a) The precursor will mix almost instantaneously with the tissue.
(b) The reactions involved are actually or formally of the first order.
(c) The reactions are effectively 'irreversible'.
(d) There is random participation of precursor molecules in the reaction.
(e) The newly formed molecules rapidly equilibrate with those previously present.
(f) There is no 'compartmentalization' of the precursor or product within the tissue.

Other assumptions in the more general use of radioactive isotopes may be added:

(g) That the 'non-specific' uptake is negligible compared with the 'specific'.
(h) That the chemistry of the radioactive isotope is the same as the non-radioactive one.
(i) That the radioactivity does not alter the rate of the reaction being studied; and
(j) that the daughter radioactive species or impurities are not present in significant number.

(a) The precursor may be accepted as mixing instantaneously in experiments *in vitro*, as it is relatively easy to arrange that the tissue be stirred rapidly. In this sense, by 'instantaneously' we mean at a rate at least two orders greater than the expected rate of the reactions we are studying. *In vivo*, this is unlikely, if not impossible, as the rate of diffusion from blood to tissue will depend largely upon the arrival of the blood at the site of uptake. This is a function of site of injection, the rate of injection, the tissue reaction with the isotope at the site of injection, the blood flow at the site of injection, the activity of the animal, the depth of anaesthesia, any stress or restraint on the animal, any tracer reaction with plasma and corpuscle components, the rate of circulation of the blood, and its rate of uptake into the target tissue itself. Although we are only interested in the latter, we cannot pretend that all the other factors are insignificant.

When a tracer is injected into an animal and a maximum concentration of radioactivity is found in the tissue, it does not mean that mixing with tissue has occurred; there may just be a high concentration of activity because of the perfusion of capillaries; the precursor may have partially reacted with particular components of the tissue. Similarly the passage of a peak of radioactivity in the vein is no proof of mixing. A possible way of testing this appears to be to freeze the part in liquid nitrogen and autoradiograph it; even then one would not necessarily know in which compartment the tracer would have to be mixed in respect of the measurement of a particular activity in order to claim that mixing had occurred.

(b) If one could be certain that the uptake of the tracer occurred along only one identified pathway, one could be certain that the reaction of that particular precursor was actually or formally of first order. When several reactions are proceeding simultaneously, but at different rates, it would be an extraordinary coincidence if analysis showed a first order reaction. If it did, it would probably indicate that the variation between samples or errors of the methods had added up to give an appearance of a first order curve. For example, in a real experiment phosphate could be incorporated into glucose phosphate, nucleotides, phosphoproteins or phospholipids, each of which may be of various kinds. Clearly, then, we must demonstrate that the reaction we are studying is by far the biggest 'consumer' of the tracer, as well as the fastest. Otherwise an apparent change in rate may be due to a redistribution of the precursor in favour of other possible pathways.

(c) The reaction need not be completely irreversible as long as the position of the equilibrium is far over to the product side. If of course the product is being continuously removed, without loss of tracer activity, the necessary condition has been attained.

(d, e) It is hoped that the precursor molecules participate randomly but demonstration of this might be very difficult. A similar uncertainty may be assumed for the equilibration of the newly-formed molecules with those already there.

(f) The assumption that there is no 'compartmentalization' of the precursor or product within the tissue very often contradicts what is believed to be the conclusion of the experiment, *i.e.* that a particular enzyme activity resides in a particular subcellular fraction. In view of the questions raised in the section on subcellular fractionation (pages 1-40), it may well be that the assumed localization is wrong or unknowable, but it would be reasonable – if without unequivocal evidence – to believe that such localization might exist.

The following other assumptions are not listed by Zilversmit et al. (1943).

(g) That 'non-specific' uptake of radioactive materials is negligible is an assumption fraught with dangers. Certainly, one cannot indulge in the luxury of saying so in the absence of the toil of finding out. The uncertainty about the single-mindedness of a reaction in an isolated enzyme preparation *in vitro* applies even more to the complex dynamic equilibria found in *in vivo*. Can one approach this knowledge? The examination of the affinity of the radioactive isotope or labelled compounds for boiled tissue, for dead tissue, and in the absence of oxygen or substrate, would seem to be urgent in order to justify the many experiments already done. Autoradiography on boiled tissue sections and other controls suggested would also be desirable (see opposite page). In autoradiography it is possible to use as a control the background counts of the number of grains in the tissue, which one is claiming not to have incorporated with the precursor (Schultze, 1969 b). One cannot use the background count rate of the microscope slide and one has to be aware that water soluble compounds, for which the tracer might have had affinity, may have been lost in earlier aqueous steps in the procedure (see page 8).

(h) The chemistry of different isotopes may not be quite the same since the isotopic mass can alter their reaction rates (Sacks, 1968). This does not seem to be important for carbon isotopes (Bigeleisen & Friedman, 1949;

Yankowich & Calvin, 1949), or sodium, potassium and phosphorus, which have larger mass numbers. However, in the case of hydrogen isotopes, the rate constants may be about 18 for tritium (Sacks, 1953).

(i) The assumption that the radioactivity does not alter the rate of the reaction being studied can only be validated for particular circumstances. The aim is always to use the minimum quantity of labelled compound with high specific activity (see page 67). Distinction should be made between the testing of the effects of whole body irradiation on the biochemistry of subsequently separated tissues, and the molecular effects of radioactive species on reactions in which they take part.

(j) Purity of the radioactive isotope in a radiochemical as distinct from a chemical sense has already been discussed (page 73). Several control experiments have already been mentioned. For tissues *in vitro*, one can use a dead, boiled, or 'inhibited' sample alongside the experimental ones.

In vivo, the following experiments are suggested:

(i) the examination of the uptake of radioactive tracers into different tissues of a dead animal, using rapid postmortem fixation by perfusion (Brown & Brierley, 1968);

(ii) the incubation of boiled homogenates and subcellular fractions with the radioactive tracers, and their subsequent complete chemical analysis to define the fate of precursors in the absence of metabolism. The measurements from both these series of experiments should be subtracted from the results of metabolic experiments;

(iii) as (ii) but with the non-radioactive compounds in physiological concentrations;

(iv) total recovery experiments for all the radioactive markers used;

(v) calibration curves with tissue present, not only on pure solutions of the non-radioactive tracer, or the compound in which its incorporation is being studied;

(vi) analytical experiments to show that the tracer is mainly and most rapidly incorporated into the pathway being studied; there is much metabolic data already available;

(vii) analytical experiments to show that the incorporation is only occurring at one identifiable site, *i.e.* not across the blood-brain barrier, the kidney, or the wall of the intestine, which are multi-component systems comprising several anatomical sites and biochemical pathways.

It is clear that control studies for experiments *in vivo* are much more difficult, but, nevertheless, should be done if we wish to analyse the results of radioactive tracer experiments meaningfully.

Arnstein, H. R., & Grant, P. T. (1957) *Prog. Biophys. Biophys. Chem.*, 7, 165.
Bigeleisen, J., & Friedman, L. (1949) *J. Chem. Phys.*, 17, 998.
Bray, H. G., & White, K. (1966) *Kinetics and Thermodynamics in Biochemistry*, 2nd edn., London, Churchill, page 222.
Broda, E. (1960) *Radioactive Isotopes in Biochemistry*, Amsterdam, Elsevier.
Broda, E., & Schonfeld, T. (1966) *The Technical Applications of Radioactivity*, Oxford, Pergamon, page 33.
Brown, A. W., & Brierley, J. B. (1968) *Brit. J. Exp. Path.*, 49, 87.
Chiewitz, O., & Hevesy, G. (1935) *Nature*, 136, 754.
Cohn, W. B. (1948) *Anal. Chem.*, 20, 498.
Francis, G. E., Mulligan, W., & Wormall, A. (1959) *Isotopic Tracers*, London, Athlone Press, page 345.
Hahn, L., Hevesy, G., & Rebbe, O. H. (1939) *Biochem. J.*, 33, 1549.
Hevesy, G. (1948) *Radioactive Indicators*, New York, Interscience, page 510.
Hevesy, G., & Hofer, E. (1934) *Nature*, 134, 879.
Laccasagne, A., & Lattes, J. S. (1924) *Compt. Rend. Acad. Sci.*, 178, 488.
Mullins, L. J. (1939) *J. Cell. Comp. Physiol.*, 14, 403.
Mullins, L. J., Fenn, W. O., Noonan, T. R., & Haege, L. (1941) *Amer. J. Physiol.*, 135, 93.
Pelc, S. R. (1951) in *Ciba Foundation Conference on Isotopes in Biochemistry*, London, Churchill, page 122.
Puck, T. T. (1960) *Prog. Biophys. & Biophys. Chem.*, 10, 237.
Reiner, J. M. (1953) *Arch. Biochem. Biophys.*, 46, 53, 80.
Sacks, J. (1953) *Isotopic Tracers in Biochemistry and Physiology*, New York, McGraw-Hill, page 5.
Sacks, J. (1968) in *Physical Techniques in Biological Research*, Vol. IIa, ed. by Moore, D. H., New York, Academic Press, page 60.
Schoenheimer, R. (1946) *Dynamic State of Body Constituents*, Cambridge, Mass., Harvard University.
Schultze, B. (1969 a, b) in *Physical Techniques in Biological Research*, Vol. IIIb, ed. by Pollister, A. W., New York, Academic Press, pages 15 and 20.
Siri, W. E. (1949) *Isotopic Tracers and Nuclear Radiations*, New York, McGraw Hill, page 388.
Skanse, B. E. (1948) *J. Clin Endocrinol*, 8, 707.
Sprinson, D. B., & Rittenberg, D. (1949) *J. Biol. Chem.*, 180, 707, 715.
Voigt, A. F. (1950) in *Biophysical Research Methods*, ed. by Uber, F. M., New York, Interscience, page 645.
Wilson, B. J. (1966) ed. *The Radiochemical Manual*, 2nd edn., Amersham, Radiochemical Centre.
Yankowitch, P. E., & Calvin, M. (1949) *J. Chem. Phys.*, 17, 109.
Zilversmit, D. B., Entenman, C., & Fisher, M. C. (1943) *J. Gen. Physiol.*, 26, 235.

5
Electrophoresis

Such separations usually consist of the following steps:

1. Killing the animal (see page 1).
2. The tissue is excised (see page 6).
3. The tissue is homogenized (see page 12).
4. The proteins, lipids, or nucleotides, are extracted.
5. The extract is placed on a gel or paper with buffer and solvents.
6. An electric current is passed through the buffer while it is being cooled externally.
7. The protein, lipid, or nucleotide, is separated, eluted, or observed.
8. The concentration of the substrate in the band or eluate is measured chemically or densitometrically.

The present examination is generalized and does not deal with the large number of specialized methods (see Henley, 1955; Bier, 1959; Moore, 1968; Bier, 1968).

4. Extraction

When the tissue is extracted each of the substances or mixtures dissolved has a particular heat of solution. This is composed of the conformational energy and the heat of solvation. Both energies are of opposite sign and are usually small for pure substances, being probably less than the -18 kcal per mol found for hydrochloric acid or -1.4 found for methanol (Gucker and Seifert, 1967). Mammalian biological tissues are less than 200 mM, but have usually been diluted at least 20 times before extraction (see page 8); the extraction is fairly slow. It would seem justifiable, therefore, to minimize the likelihood of a significant temperature change.

5. Extract placed with solvents and buffer

The extract is placed with solvents and buffer on the gel, paper, or near an electrode. Due to the requirement that it should carry a current, the solvent is usually aqueous. The effectiveness of different solvents, buffers, pHs, and concentrations for the separation of particular compounds, indicates that the materials being separated must have some kind of physico-chemical relationship with the medium.

6. Current is passed through buffer

In electrophoresis the biggest single problem is heat dissipation (see Bier, 1968). It is normally assumed that the cooling prevents the temperature rise. Although successful electrophoresis could not be carried out until the heat problem was recognized (Tiselius, 1937), it is unusual to calculate the heat generated. While cooling is being applied the total heat generated will *always* be given by $Vit/4.2$ calories, where V is the voltage, i is the current and t is the time in seconds. As progressively higher voltages are used, the electric power may become considerable.

The heat generated in each region separated will depend on:

(a) the total voltage applied between the electrodes;
(b) the voltage drop across a particular band;
(c) the resistance of the electrodes in series with the preparation;
(d) the overall resistance of the electrophoretic strip;
(e) the specific resistance of each of the materials separating at that moment in the band, and at the precise temperature and current at the moment within it.
(f) the specific resistance of the gel or paper and buffer 'in parallel' with the material being separated; this itself is a function of the total chemistry of the medium in which the electrophoresis is being carried out;
(g) the relative proportions of good and poor conducting materials in the system, *e.g.* the buffer and paper, respectively; and
(h) the duration of the electrophoresis.

The thinner the glass, the more rapid the cooling, and, therefore, the less the temperature will rise. Nevertheless, it must be insisted that no cooling system *whatsoever* will prevent the heat being generated when the current passes. The hope is that the heat will be dissipated *before* the temperature can rise. It is assumed that the substances being separated are so small that they can be treated as a few molecules to which energy added as a result of the passage of electric current, or friction, will be distributed randomly. In view of the fact that the extracts are usually in solution in water or organic solvents, this assumption is probably permissible.

There has sometimes been a temptation to use higher and higher voltages because 'better' separations have been achieved by them. The question arising in respect of large molecules such as proteins and lipids, for example, is how we can distinguish between the separation of several original components of what was formerly believed to be a homogeneous system and the degradation into several unnatural products of a compound which was homogeneous in the native state. There may be no absolute solution to this problem but three experimental attitudes are perhaps relevant. Firstly, we cannot regard the separation of more and more peaks as evidence in itself that the parent compound was not 'pure' — that is, the multiplication of electrophoretic peaks need not be regarded as an aim in itself. Secondly, 'purity' is not necessarily a relevant aim in biochemistry; the understanding of the relations between compounds in a 'native' state will probably yield more valuable information about the chemistry of living processes. Thirdly, we should try to test for 'functional' properties, like enzyme activities, or 'nerve growth factor' potential, or hormonal activity, as measures of 'purity', if we can satisfy ourselves that these themselves have not been substantially altered by separation techniques involving the injection of the minimal quantity of energy into the system (see page 108).

We can easily calculate the total heat generated during electrophoresis. If we take, for example, the separation of serum proteins by paper electrophoresis, Wunderly (1959) gives a table of voltages, currents and times used by five different groups of workers. These voltages varied from 110 V to 340 V, and current intensities varied from 0.4 mA to 7 mA; separations were carried out for between 3 hours and 17 hours. Since the 1950s when these experiments were done, the tendency has been to use higher voltages for shorter times. Nevertheless, we can calculate that the heat generated in these conditions was between 250 calories and 6900 calories during the whole period of electrophoresis. We cannot know whether or not to be alarmed by the injection of these quantities of heat into such notoriously temperature sensitive biochemicals as proteins, without knowing the size of the extract placed on the strip and what proportion of it will be dissipated in the fractions which interest us. The factors upon which the proportions depend are listed in (b) to (g), and will be dealt with in more detail below.

In a particular band the heat generated will depend upon the voltage gradient across itself, and its specific resistance. As it becomes 'purer,' its specific resistance will be changing all the time, as it will also during its migration along the voltage gradient. Since the overall voltage between the electrodes is held relatively constant, the heat generated in each band must be changing continuously and gradually. It will achieve a steady state when the band has stopped moving and when these two parameters upon which it is dependent have become constant; this may happen at different times for different bands.

The resistance of the electrode will change even although there is a 'salt bridge'. It is often assumed — somewhat naïvely it seems to the author — that having a long salt bridge means that the electrodes will not be oxidized or reduced, *i.e.* that Faraday's Law will disappear from the chemical statue book. However, the main change in electrode resistance probably occurs early on when the electrophoresis is 'settling down', as the overall voltage and current as measured are held constant through a separation.

The overall resistance between the electrodes will determine the total current passing but much will pass in parallel with our compounds through the buffer, the salts, the gel or the paper. It should be remembered that the gel or paper may have a very high resistance and therefore most of the current would pass through the buffer solutions. In each band the heat generated will depend upon the ratio of the resistance of that band compared with the resistance of the parallel channels. This will itself depend to some extent on the temperature of the different constituents of the system. As the temperature rises, the resistances of each component will go down according to the temperature coefficients of each of the materials within the system; it is extraordinarily unlikely either that these coefficients would be the same. As far as the author is aware, these are not known for the conditions of electrophoresis. It is likely – and it could and should be tested – that if one were to do the control experiment of carrying out an electrophoretic run on the solvent system plus the buffer, but without the biological extract itself, the resistance would not be measurably different than in the presence of the biological extract; this would indicate that the very small quantity of extract had a very low resistance, and, therefore, little heat would be generated in it, compared with the rest of the system. This would not, of course, alter the fact that in the voltage gradient different amounts of heat would be generated in the region of different compounds being separated due to their different specific resistances.

If one is carrying out an electrophoresis on paper, or other material of high specific resistance, nearly all the current will pass through the buffer. Therefore, the heat generated in the band will depend upon the relative volumes of the high and low resistance materials within the system.

Obviously, the total heat generated will depend upon the time the electrophoresis takes. The heat dissipation will occur at a finite rate; therefore, it is obviously more desirable to use the smallest current for the longest time, which will generate the same total quantity of heat, but permit adequate heat dissipation before the temperature has risen. However, slow electrophoresis permits diffusion and convection which may diminish the resolution of the system.

It is, of course, the temperature rise in the preparation which will affect the enzymes and other proteins. This will depend upon all the factors listed above, as well the *rate of dissipation* of the heat. Perhaps it should not be necessary to reiterate that however rapid the cooling it will have no effect on the rate of heat generation. We are closing the stable door after the horse has bolted, unless the horse fails to rise from his haunches with sufficient alacrity.

The temperature rise will also depend upon:

(a) the *quantity* and *rate* of heat generation, due to the factors listed above;

(b) the efficiency of the cooling system;
(c) the distance of the coolant from a particular band;
(d) the absolute and relative heat conductivities of the band, the extracting solution, the electrophoretic medium, and the glass;
(e) the total quantity of material in each band; and
(f) the total geometry of the electrophoretic strip.

Critical parameters are the heat conductivities and capacities of the materials in the different peaks and the media surrounding them. Since each material would be very likely to be different, the temperature rises would be very likely to be different as they would depend on the *rate of heat dissipation* between the biological extract, the buffer, and the material of the apparatus.

If one were to start out with a mixture containing enzymes of the same activity but different specific resistances and mobilities, they would be 'denatured' to different extents and thus they would end up appearing to have different enzymic activities. If the energy is sufficient to break down a compound, the resulting degradation products would almost certainly have different electrophoretic mobilities.

Thus it would seem that there is a case for measuring the heat conductivities of proteins, lipids and nucleotides in conditions simulating an electrophoretic separation. Similarly the specific resistance of the bands should be measured. If it were found that the latter was negligible compared with the medium, the heat generated could be calculated from the position of the band within the overall voltage gradient between the electrodes. The temperature rise could thus also be calculated. Failing this, if one were to calibrate protein solutions for denaturation due to known temperature rises, and then measure the denaturation within the electrophoretic strip, some idea would be obtained of its magnitude. Unfortunately, it is not enough to measure the temperature 'within the apparatus'.

7. Separation, elution or observation

Many modern electrophoretic techniques separate the fractions by convection or decantation. Some require elution and sometimes optical techniques are used instead of elution. Separation in electrical or magnetic fields involves the generation of heat. Elution is an extraction procedure in which the loose chemical or physical bonds between the substance and the medium are broken (see the discussion on chromatography, page 89); it requires a high percentage extraction from each part of the gel, paper, powder, or resin, and is only quantitative if this condition is fulfilled.

8. Chemical or densitometric measurement of substrate concentration

The concentrations of the separated substances are measured by classical chemical or densitometric techniques. Examination of moving boundaries or bands, optically and microscopically, has received a great deal of attention both theoretically and practically (Tiselius, 1937; Longsworth, 1959; Lerche, 1953; Brinton & Laufer, 1959). Only a few brief remarks need be made.

Measurements are made of the intensity, refractive index, absorption or other optical functions of the substances being separated. The accuracy of the measurements will depend upon:

(a) the system of electrophoresis;
(b) the homogeneity of the band, both physically and chemically;
(c) the difference in refractive indices between the bands and the medium;
(d) heat generation causing convection currents;
(e) impurities in the chamber;
(f) the geometry of the chamber; and
(g) the accuracy of alignment of the optical system.

Assumptions of electrophoretic techniques

(a) All the assumptions appropriate to chromatography (see page 92).
(b) That the heat generated by the electrical or magnetic fields, will not be dissipated in the preparation being separated.*

---

* this assumption contradicts the laws of physics.

(c) That cooling prevents the temperature rising significantly (see page 15).
(d) That clear separation of peaks implies that the compounds *in vivo* were distinct, *i.e.* that bonds have not been broken by the procedure.
(e) That polymerization or acceleration of reactions between compounds not normally associated does not occur.
(f) That separation of a biological activity in a peak implies that the biological activity has not been decreased by the separation.

Assumptions (a) derive from the fact that most of the measures used in chromatography (*q.v.*) are also employed in electrophoresis.

The heat generated by electrophoresis has been discussed at length (page 81). Unlike the cases of subcellular fractionation or electronmicroscopy, its exact quantity and power can be calculated. The effect of cooling is to increase the resistance and this will be reflected in a change of current. In assessing the heat generated, one must also take into account the initial 'settling down' period when the apparatus is turned on.

The important distinction between heat being generated and the temperature rise resulting from it has been frequently urged (see page 16). The question which may be unanswerable directly is whether the temperature rise at the particle surface is sufficient to 'denature' proteins; in other words, is there a temperature gradient in the bulk solution and, or, in the parts of the extract? As in chromatography, it is very often a solution and the energy is likely to be distributed randomly. There is, however, a further indication that electrophoresis is not too dangerous. Biological activities are some of the most unstable of properties, yet tetanus toxin and diphtheria toxoid (Largier, 1957), diphtheria anti-toxin (Cann, Kirkwood, Brown & Plescia, 1949), poliomyelitis virus (Polson, 1953), phosphocreatine and nucleotides (Wollenberger, 1954), antibiotics (King & Doery, 1953), and many other extremely unstable materials, seem to survive electrophoresis. However, when recoveries have been measured they have not always been 100 per cent (see, for example, Polson, 1953; Bier, 1957).

One should make this distinction between survival of the biological or biochemical activity, with an unknown recovery, which is a normal commercial or clinical requirement, and a measured high percentage recovery of activity in components of the system to be compared, *i.e.* assumption (f). From the quantitative biochemical point of view, we are interested in the latter assay.

Assumption (d), that the more peaks on electrophoresis, the more separate chemical compounds there were originally, is pleasing but uncertain. It is, nevertheless, an aim of many biochemists. Although we cannot know whether a compound has been degraded or whether new inter-reactions have occurred in the preparation (e), we can calculate the power of the electrophoresis, and systematically examine to see if that amount of power does alter the chemistry of the system. Although such an experiment would be empirical, it would seem necessary for the consideration of the applicability of electrophoresis for examination of unstable compounds.

At this stage we may be permitted to specify desirable characteristics for electrophoretic separations — that they be carried out with the lowest possible current in the longest possible time, and with the maximum rate of cooling; that they are not used in any temperature-sensitive system without measuring the effect of the temperature rise. It would also be desirable to plot the heat generated at particular spots against the mobility of the substances being separated, their degree of 'denaturation', and their enzyme activities, as well as other temperature sensitive properties. Relatively simple relationships might exist. It is also pertinent to wonder whether as many proteins or isoenzymes would be detected as hitherto.

Control experiments for electrophoretic techniques

(i) Measurement of the loss of biological and chemical activity of temperature-sensitive substances.
(ii) Recovery experiments for the whole electrophoretic system.
(iii) Examination of the movements of co-factors in the electrophoretic field, as their migration would affect the enzyme activity detected.

Furthermore, verbal discussions of how much these factors may affect our measurements in biochemistry can be of little value in the absence of these simple but fundamental experimental controls. These would either satisfy us that such effects would be too small to affect our results, or large enough that we must correct for them, or, more likely, that they yield valuable information on the native states of the compounds.

Bier, M. (1957) *Science*, 125, 157.
Bier, M. (1959) *Electrophoresis, Theory, Methods, and Applications*, New York, Academic Press.
Bier, M. (1968) in *Methods in Enzymology*, Vol. V, New York, Academic Press, page 33.
Brinton, Jr., C. B., & Laufer, M. A. (1959) in Bier, M. (1959) page 427, *op. cit.*
Cann, J. R., Kirkwood, J. G., Brown, R. A., & Prescia, O. J. (1949) *J. Am. Chem. Soc.*, 71, 1603.
Gucker, F. T., & Seifert, R. L. (1967) *Physical Chemistry*, London, E.U.P., page 280.
Henley, A. (1955) *Electrophoresis Bibliography*, Maryland, American Instrument Company.
King, N. Kelso, & Doery, H. M. (1953) *Nature*, 171, 878.
Largier, J. F. (1957) *J. Immunol.*, 79, 181.
Lerche, C. (1953) *Acta. Pathol. Microbiol. Scand.*, Suppl. 98, 1.
Longsworth, L. G. (1959) in Bier M. (1959) page 142, *op. cit.*
Moore, D. H. (1968) in *Physical Techniques in Biological Research*, Vol. IIIa, *Physical Chemical Techniques*, ed. by Moore, D. H., New York, Academic Press, page 121.
Polson, A. (1953) *Biochem. Biophys. Acta.*, 11, 315.
Tiselius, A. (1937) *Trans. Farad. Soc.*, 33, 524.
Wollenberger, A. (1954) *Nature*, 173, 205.
Wunderly, Ch. (1959) in Bier, M. (1959) page 184, *op. cit.*

6

Chromatography

Chromatography consists of the following stages:

1. Killing the animal (page 1).
2. The dead animal cools (page 5).
3. The tissue is diluted (page 8).
4. The tissue is homogenized (page 12).
5. An extract of the lipids, carbohydrates, or amino-acids is made (page 79).
6. The extract is buffered, and then added to a column, paper or liquid, or put into a gas flow.
7. Solvents are passed in one or two dimensions.
8. Paper is dried.
9. The column is eluted, or the paper is developed.
10. The quantity of the material is measured chemically, densitometrically, isotopically, or by thermal conductivity, ionization, or electrical conductivity techniques.

Early reviews listing the techniques and assumptions are of great interest (see, for example, Tswett, 1910; Zechmeister & Cholnoky, 1941), and there are also several excellent modern ones (Lederer & Lederer, 1961; Heftmann, 1961; Burchfield & Storrs, 1962; Hais & Macek, 1963; James & Morris, 1964; Porter, 1969; Smith, 1969).

6. Extract is buffered and added to column or paper

The extract is buffered and added to the column or paper, on which it is absorbed (Lederer & Lederer, 1961, page 19). Different kinds of column absorb different materials from extracts (Strain, 1942; Le Rosen, Monaghan, Rivet & Smith, 1951) and the absorption also depends on the pH and temperature. Therefore, the different compounds being separated must react physico-chemically with the materials to different degrees. Different amounts of heat will be released by reaction with the absorbent (Muller, 1943). The temperature rise will depend upon a great many factors (see page 15). Paper and gas chromatographs are relatively open systems compared with columns.

7. Solvents are passed in one or two dimensions

The rate of migration will depend upon: the quantity and composition of the material being separated; the nature, concentration, pH, temperature, and amount of the mobile phase; the particle size, nature and packing of the stationery phase (Ambrose, *et al.*, 1960); the pressure on the column; the absolute and relative volumes of the extract and the column; the previous treatment of the column; the number of 'washes' (Martin & Synge, 1941; Cassidy, 1957). Nearly all these properties change continuously during the experiment. The native activating or inhibiting cations or other co-factors may migrate at different rates than the enzymes with which they are normally associated. When these co-factors are known the columns are usually saturated with them or deprived of them, as the experiment requires. It would, however, be more desirable to add a concentration of these substances, which is close to that *in vivo*. One could use either 'extracellular' or 'intracellular' fluid, (see page 112).

In gas chromatography the characteristics of the carrier phase, which affect the separation, have been identified (for discussion see Burchfield & Storrs, 1962, page 94). They are the molecular weight of the gas, its diffusion coefficients into hydrogen and carbon dioxide, its viscosity, thermal conductivity and ionization potential. These examinations could, and should, be extended to the physical properties of the solvents normally used in column and paper chromatography. At present the choice of solvents is usually made by reference to previous publications by authors who have arrived at their techniques empirically. Ideally one should work towards a situation in which one could arrange solvents in a series of tables according to their physico-chemical properties and thus be able to choose those most likely to separate a particular species from a mixture. Theoretical or empirical laws could probably be enunciated. Furthermore, the nature of each solvent would give important information about the bonds connecting the material to be extracted with other native chemicals, (see page 31).

We can specify desirable characteristics of solvents for chromatography. First, they should separate the compounds in a mixture as sharply as possible, and, secondly, they should have been demonstrated not to alter the enzyme activities significantly. Since we measure enzyme concentrations by their activities, we must always beware lest the solvents perform their own inhibitions, activations or denaturations. In this respect also, we might ask if there is a process analogous to denaturation of proteins which occurs in nucleotides, carbohydrates and fats.

8. The paper is dried

Drying the paper fixes the spots in particular positions, just as the chemical reactions fix particular components in particular regions of a column. Drying has the important advantage that it extracts heat from the sample and thus minimizes changes during this stage of the procedure. Drying produces a change in conformation of any molecules containing water. We know that the subtle chemistry of peas must be changed irreversibly when they are dehydrated, as we can detect a difference in taste between fresh peas and peas which have been dried. However, spores, antibiotics, sperm and some toxins, are not necessarily changed irreversibly with respect to their biological activity (Smith, 1960; Wolstenholme & O'Connor, 1970) so that in separating large molecules by electrophoresis or chromatography, the reversibility of any change in activity on drying should be measured. It is reasonable to suppose that many of the effects of freezing tissue below the eutectic point of the salts in it are due to dehydration.

9. The column is eluted or the paper developed

In order to elute the separated compounds off the column, one may use the same or a different series of solvents and conditions than those which encouraged the association with the column. This raises two questions. Firstly, what is the recovery of enzyme activity when *added* to a column, and, secondly, what is the recovery following elution from that column, with or without washings? Just as heat will be generated when the compounds are added to the column, it may be liberated or absorbed when they are desorbed or eluted off it. Fortunately, we can measure both recoveries added together by comparing enzyme activities in a known mixture before and after chromatography. It cannot be assumed that recovery of each compound will be either similar or total. The usual criterion cited is sharpness of peak rather than recovery of activity. When recoveries are done the recovery of a particular compound is sometimes expressed as a percentage of the sum of the activity surviving the column, rather than the activity of the extract initially put onto the column.

10. Measurements

Chemical, densitometric, isotopic, thermal conductivity, ionization or electrical conductivity measurements are made on paper or columns. When paper is 'developed' with ninhydrin or amidoschwartz or other developers (Smith, 1969), the procedure is not usually quantitative, unless it be assumed that these substances react equally with each of the species examined. This problem has been tackled directly in autoanalytical measurements of aminoacids by the use of leucine equivalents (Moore & Stein, 1954); this principle could be modified for recovery from all chromatography procedures.

The relative efficiencies of different detecting techniques used in gas chromatography have been reviewed by Hill, (1969). An important symposium on quantitative paper and thin layer chromatography contains several papers on measurement and detection problems, particularly optical ones (Shellard, 1968). It is worth pointing out that chromatography probably involves less heat generation and temperature rise than centrifugation, electronmicroscopy or electrophoresis.

Assumptions implied in the use of chromatography

(a) All those assumptions appropriate to killing the animal, homogenizing the tissue, and extracting the mixture to be chromatographed.

(b) Adding the extract in a solvent system to the paper or column does not induce significant temperature rise.†
(c) That the same amount of heat would be generated at different points between the origin and the solvent front.*
(d) That 'washing' the column causes no significant loss of the compound or change in its reactivity with the column.†
(e) That drying the spot on the paper does not cause significant irreversible change in its chemistry, *e.g.* by polymerization, crystallization or dehydration.
(f) That a second solvent system has no effect on the chemistry of the separated compounds.
(g) That desorption or elution of the bands or spots with other solvents does not cause significant temperature rise.
(h) That the measuring system does not degrade the compounds in the bands or spots.

It would be very difficult to measure the temperature rise at the surface of the extract on addition of the material to the column or paper, but some idea of the order of the quantity of heat generated can be calculated (see, for example, Muller, 1943). Although mention is made of exothermic reactions, the real danger is that the different materials will be affected to different extents by the change in free energy of each of the separate reactions.

Assumption (c) that the same amount of heat will be generated is not tenable because the heat will be proportional to the friction with the medium and the distance a material has travelled.

Whether or not there is significant loss of activity in a column or on paper – see assumption (d) – is not always measured. This does not necessarily invalidate the procedure if the separation of a mixture is an end in itself; for example, when it is used clinically. When, however, we are attempting to make quantitative biochemical statements, recovery experiments are obligatory.

'Washing' a column is often used to dilute away unwanted or 'contaminating' compounds similarly mobile but less firmly

---

* this assumption contradicts laws of thermodynamics of physics.
† in some experimental systems, this has been shown to be untrue.

bound. Again, here, we should distinguish between our desire to end up with a chemically 'pure' material and our belief that all of it has survived.

Chromatography is used as a separation procedure for adenosine polyphosphates (Siekevitz & Potter, 1953), phosphocreatine (Heald, 1960), arginine phosphate (Ennor & Rosenberg, 1952), antibiotics (see Jones, 1954), 'nerve growth factor' (see Levi-Montalcini & Angeletti, 1968), adrenalin (Bergstrom & Hansson, 1951), and noradrenalin (Bergstrom & Sjovall, 1951), among many other extremely labile substances (see Lederer & Lederer, 1961). This does not mean that other techniques necessarily degrade these biologically active molecules or that chromatography does not change their activity, but its popularity for isolating unstable materials still seems to be increasing and this in itself indicates its effectiveness in separating unstable materials without excessively degrading them.

Control Experiments

(i) All the controls appropriate to the first four steps mentioned previously.

(ii) Comparison of the activity of a known purified enzyme at each stage of the chromatographic procedure, and overall recovery of the whole process.

(iii) As (i), but with the addition of excess known purified enzyme to a homogenate.

(iv) Systematic study of the chemical reactions of paper and column materials together with the substance being separated.

(v) Examination of the effect of the solvents and the eluting solutions on the enzyme activities.

Ambrose, D., James, A. T., Keulemans, A. I. M., Kovants, E., Rock, R., Rouit, C., & Stross, F. H. (1960) *Pure and Appl. Chem.*, 1, 177.
Bergstrom, S., & Hansson, G. (1951) *Acta. Physiol. Scand.*, 22, 87.
Bergstrom, S., & Sjovall, J., (1951) *Acta. Physiol. Scand.*, 23, 91.
Burchfield, H. P., & Storrs, E. E. (1962) *Biochemical Applications of Gas Chromatography*, New York, Academic Press.
Cassidy, H. G. (1957) *Fundamentals of Chromatography*, New York, Interscience, page 29.
Ennor, A. H., & Rosenberg, H. (1952) *Biochem. J.*, 51, 606.
Hais, I. M., & Macek, K. (1963) *Paper Chromatography*, Prague, Czechoslovak Academy of Sciences.

Heald, P. J. (1960) *Phosphorus Metabolism of the Brain*, Oxford, Pergamon.
Heftmann, E. (1961) *Chromatography*, New York, Reinhold.
Hill, D. W. (1969) *Detectors for Gas Chromatography*, in Porter, R. (1969), *op. cit.* page 37.
James, A. T., & Morris, L. J. (1964) *New Biochemical Separations*, London, Van Nostrand.
Jones, T. S. G. (1954) *Brit. Med. Bull.,* 10, 224.
Lederer, E., & Lederer, M. (1961) *Chromatography*, Amsterdam, Elsevier.
Le Rosen, A. L., Monaghan, P. H., Rivet, C. A., & Smith, E. D. (1951) *Anal. Chem.,* 23, 730.
Levi-Montalcini, R., & Angeletti, P. U. (1968) in *Growth of the Nervous System Ciba Symposium*, London, Churchill, page 126.
Martin, A. J. P., & Synge, R. L. M. (1941) *Biochem. J.,* 35, 1358.
Moore, S., & Stein, W. H. (1954) *J. Biol. Chem.,* 211, 893.
Muller, P. B. (1943) *Helv. Chim. Acta,* 26, 1945.
Porter, R. (1969) *Gas Chromatography in Biology and Medicine*, Ciba Symposium, London, Churchill.
Shelland, E. J. (1969) *Quantitative Paper and Thin-layer Chromatography*, London, Academic Press.
Siekevitz, P., & Potter, V. R. (1953) *J. Biol. Chem.,* 201, 1.
Smith, A. U. (1960) *Biological Effects of Freezing and Supercooling*, London, Arnold.
Smith, I. (1969) *Chromatographic and Electrophoretic Techniques*, 3rd edn., London, Heinemann.
Strain, H. H. (1942) *Chromatographic Adsorption Analysis*, New York, Interscience.
Tswett, M. (1910) *Chromophylls in the Plant and Animal World*, Warsaw, Tipogr. Warshawskogo Utshebnago Okruga.
Wolstenholme, G. E. W., & O'Connor, M. (1970) *The Frozen Cell*, Ciba Foundation Symposium, London, Churchill.
Zechmeister, L., & Cholnoky, L. (1941) *Principles and Practice of Chromatography*, London, Chapman & Hall.

7
General characteristics of all techniques considered

An animal is taken from its natural environment where it may have lived under a less than optimal regime, or from the artificial but more charitable environment of an animal house. It is subjected to a certain amount of stress or pain; some of this occurs with killing, though much can be avoided.

Following its death, the animal cools, and autolysis supervenes. The impermeability to many substances diminishes, resistance to infection disappears and previously unevenly distributed substances travel along their concentration gradients. Tissue is excised at various times afterwards and subjected to strong fixatives, organic solvents or acids, all of which are known to have powerful biochemical effects which are generally ignored. Considerable dilutions, denaturations, inhibitions, oxidations, etc., occur during the addition of these agents. The tissue is then sliced or homogenized; it may then be centrifuged. Homogenization and centrifugation both generate heat in a system which we would like to be open but may be partly isolated.

Further substrates — with or without stains or extractants — are then added, followed by another cohort of inimical chemicals. Each has its own heat of solution, dilution or reaction, which may be positive or negative. Absorption onto a column or paper, or desorbtion from it, generates heat as does electrophoresis. Each separate constituent raises its temperature differently so we must again hope that the system approximates more to an open than a closed one. Products of the reactions are extracted in other exergonic or endergonic reactions. The measurements are usually calibrated with pure solutions in the belief that the intensity or activity of the chromatophore has not been altered by the whole separation procedure.

Using this generalized description, what are the minimum

General characteristics of all techniques considered 97

hiatuses in our knowledge which must be filled before we can make meaningful statements which may reflect quantitatively the state and concentrations of labile materials in the tissue *in vivo*? These may be listed:

1. the effects of killing;
2. the biochemical changes occurring in the agonal state;
3. the effects of cooling;
4. the effects of the media in which homogenization is being carried out;
5. the effects of dilution;
6. the effects of homogenization;
7. the effects on the activity we are studying of all fixatives, extractants, precipitates, adsorbents, etc., used at any time during the procedures;
8. the partition coefficients between different phases of the preparation of any of the above reagents;
9. the rate of temperature rise or heat generation of any procedure;
10. the reversibility of any effect of any reagent following its subsequent removal;
11. the study of tissue levels *in vivo* of known co-factors;
12. the heat conductivity of different tissues, fractions and particles;
13. if experiments 4 to 12 cannot be carried out, the recovery of a purified enzyme or protein of known initial activity could be measured at each stage;
14. the lability of common macromolecules to temperature rises occurring at the particular rates which have been found to result from each technique whose use is contemplated;
15. the arrangement of techniques in a hierarchy in respect of their optimal use for a particular measurement planned (see page 105).

It will be noticed that all these series of experiments are empirical, as biological tissues are complicated from a chemical point of view and often not in equilibrium. I believe that until the validity of our techniques for studying biological material has been established quantitatively, the reliability of physico-chemical interpretations of biochemistry will be correspondingly limited.

8

The individual techniques

Subcellular fractionation

So many and unlikely are the assumptions involved in subcellular fractionation (page 33) that it is highly unlikely that comparisons of chemical activities between different fractions have any value at all. In the normally accepted procedures, different amounts of work are done in different parts of the centrifuge tubes and different fractions to be compared are centrifuged for different times with added co-factors. The only possible way the experimental results could be justified would be to show that the effects of the procedures were extremely small compared with the parameters which were being measured. These demonstrations have yet to be made.

We should perhaps distinguish between two different kinds of ignorance about subcellular fractions. Firstly, there are the endoplasmic reticula, sarcoplasmic reticula, nuclear pores, etc., which must be artefacts (see page 59). Secondly, there are the nuclei, the mitochondria, and the cell membranes, which can be seen by light microscopy; it would seem quite possible that quantitatively we know very little about their relative biochemical activities. This means that we cannot assess the meaning of any sentence starting, 'mitochondrial fractions' or 'ribosomes' or 'membrane fractions' etc.

It is often said that subcellular fractionation techniques have greatly expanded our knowledge of biochemistry. If all the quantitative information is uncertain, what phenomena which were not previously known in crude homogenates can be regarded as having been discovered by these methods?

Histochemistry

This technique is at its best only descriptive. It cannot be quantitative except in a comparative way because it cannot be calibrated. If one finds an enzyme activity at a particular site one can accept that a certain but unknown proportion of that originally present at that site has survived there or has come from elsewhere; if the enzyme is soluble it may only be a very small proportion. If, on the other hand, one does not detect an expected enzyme, we cannot know whether or not the substrate, co-factors or indicator have diffused away, or, indeed, have any affinity with the site. The observation that enzymes detected histochemically are usually seen on membranes is not without implications.

Electronmicroscopy

This is such a drastic procedure that the new information which has been found by it is also very limited. It is a technique examining the leather rather than the skin (see page 103). Because of our ignorance about the complex chemistry and precipitating properties of chemical constituents of living tissue, we cannot know whether any structure seen is an artefact or not, unless it has previously been observed by light microscopy. This is especially true for the cytoplasm which is a suspension *in vivo* to which we add non-miscible organic solvents.

The relative shrinkage of different parts of cells — which has not been measured — does not permit us to make any measurements of the relative sizes of organelles or the spaces between them. Nor can we know the shapes of particles, since the osmium salts are a shadow and the intense radiation probably makes particles tend to regain their minimum diameters.

With electronmicroscopy, unlike subcellular fractionation, most of the artefacts and difficulties have been fully described and research workers discuss them publicly. Yet they continue to carry out the techniques as if such limitations had never been demonstrated.

Electronmicroscopy is a very clear example of an uncertain technique. It is interesting that it was developed for the study of very stable inorganic materials and then taken over by the biologists. In order to examine tissue we must fix, dehydrate and irradiate it. If any part of it is significantly sensitive to *any* of these steps, we will never see it properly. Yet we have plenty of information about the sensitivity.

The usefulness of the technique is mainly in demonstrating osmiophilic structures without indication of their relative sizes or shapes. It can only give information about labile materials within the system if they have been firmly bound to the tissue before preparation. Even here its usefulness is dependent upon the staining material not affecting the labile activity.

Electronmicroscopy can also be used to compare normal and abnormal tissue from the same organ but usually the resultant description is only a verbal one. There is pantheon of bodies with particular shapes – Golgi bodies, vesicular bodies – and a family of somes of which Galsworthy would have been proud. They have been given living characteristics. Synaptic vesicles have been filled with acetylcholine, although it has been calculated that the amount of acetylcholine liberated during electrical activity would require them to be solid acetylcholine. Lysozomes have been denigrated as 'suicide bags' and theories have been adumbrated to show how they avoid catabolizing themselves.

Electronmicroscopy was developed for metallurgical purposes. Here the substances studied are inorganic, stable and usually crystalline. Biological tissue is relatively unstable, aqueous and heterogenous.

We may summarize conclusions about subcellular fractionation, histochemistry, and electronmicroscopy by saying that they may be useful to compare normal and abnormal tissue. They may also contribute to histology. What little information histochemistry gives us of biochemical relevance is only about the water-insoluble organelles.

Radioactive measurements

These techniques have made important contributions to our knowledge of metabolic pathways. They have a significant advantage in that they are relatively non-energetic and thus do not disturb the tissues which they are being used to examine. However, both *in vivo* and *in vitro*, they are not by any means as specific as their protagonists would like them to be, mainly due to the several assumptions which their use implies. *In vivo* it is often difficult to do the necessary control experiments and, therefore, isotopes are far more useful for *detecting* metabolic pathways than in measuring the rates of reactions within them.

Many of the alleged quantitative results of experiments have ignored the assumptions listed and understood by the pioneers (see page 74). *In vitro*, the control experiments are much easier (see page 70) but there are many extra problems related to the unusual environment of the tissue.

The prime advantage of radioactive measurements is that they may avoid the necessity for killing the animal, taking out the tissue and extracting it. They can be used in dynamic studies with several observations on the same animal or acute killing and counting of a whole organ. In many senses, radioactive techniques are the least uncertain. They can sometimes be made quantitative even *in vivo*.

Electrophoresis

As has been pointed out, this technique is used successfully to separate various biological materials (page 86). The heat generated within the system can be assessed and it is fairly high, although spread over a long period. The cooling and the smallness of the quantity of the extract applied are evidently both effective in preventing excessive temperature rise, otherwise labile biological materials would lose their activity. The measure of recovery of a labile activity is the best criterion for the suitability of a technique for biological studies. Nevertheless, complete recoveries of all isoenzymes, serum proteins, etc., should be done, as there is undoubtedly a considerable heat gradient along the electrophoretic strip.

Electrophoresis is an example of one technique in which the suggested parameters of rate of rise of heat generation and the constant level of heat generation, could easily be indicated.

Chromatography

This is probably the 'mildest' technique involving extraction. It is often used for separating fairly small molecules. Heat is generated during the absorption or desorption but the quantities involved are extremely small and the system approximates very closely to an open one. However, we should recollect that all the other steps in the procedure have already been taken (see page 89) and we are adding a minute extract to the column or paper. It would be more desirable to develop those techniques in which a *piece of tissue* is placed on the column and the minimum number of solvents extract and separate it at the same time.

A considerable advantage of chromatography is that it is comparatively easy to use a known quantity of a known substance both as a marker and to measure recovery.

9

The definition of biochemistry

In order to discuss this in a useful way we must first ask what is the purpose of a particular technique. The more fundamental question is, what does the term biochemistry mean? We can study (a) a preparative technique for a biologically active compound; (b) the tissue level of a biochemical constituent; (c) the chemistry of a material of biological origin, *e.g.* leather, wool or wood; or (d) the chemistry of metabolizing tissue *in vivo* or *in vitro*, like skin, hair, or xylem.

Preparative techniques have the virtue that a commercial interest often ensures a high yield so that recovery experiments are often done. We must distinguish here between the purity – as evinced by the sharpness of a chromatographic peak, for example – and the recovery or concentration of most of the initial activity of *all* the components in the peaks being compared, both with each other and with peaks from preparations in control conditions (see page 119).

The tissue levels of biological constituents, such as serum proteins, urinary steroids, cerebrospinal fluid or antigens, are usually measured for diagnostic or prognostic clinical purposes; biopsy specimens are also taken for histology or biochemical analysis with the same intention. The measurements made on these specimens and their relationships with disease states are observed statistically. Such measurements are extremely valuable in the sense that because they are often empirically established, they are relatively free of the sort of theoretical assumptions discussed here (see pages 33, 49, 63, 73, 85, 92). By the same token, the normal values are found by observations on healthy subjects. However, their extreme usefulness clinically should not blind us to the proposition that they probably bear little relationship to the tissue levels *in vivo* of these constituents. For example, between the presence of clinical levels of potassium or sodium ions in the serum of a patient and measurement of them by flame photometry, the following events will have occurred to influence the relative levels between serum and cells:

(i) the blood has been withdrawn through a fine needle;
(ii) it has cooled down;
(iii) it has been injected into a centrifuge tube;
(iv) it has been broken into a turbulent fluid;
(v) the cells have continued to metabolize;
(vi) the partial pressure of the oxygen in the fluid away from the surface has decreased;
(vii) the carbon dioxide and lactic acid concentrations in the depth of the solution have increased;
(viii) the sample has been centrifuged, and the serum has been pipetted off.

Each of these steps will have the effect of increasing the passive diffusion of potassium ions out of the red cells into the serum and of sodium ions in the opposite direction. Published values for these constituents in the serum will consistently overestimate the potassium ion and underestimate the sodium ion concentration. Nevertheless, the clinicians are still in an unassailable position as their normal samples have been treated in the same way. If one wanted to extrapolate back to the real tissue values *in vivo* one would have to evaluate systematically the effects of each step on the parameters one was studying; one could then try to minimize them technically or correct for them mathematically.

In experimental research, on the other hand, one should make the very clear distinction between the two attitudes covered by the term 'biochemistry'; the one in (c) being the chemistry of tissue of biological origin — the leather — and (d) the chemistry of biological processes themselves — the living skin. I would submit that many of the biochemical techniques being used today so alter the tissue, that one is looking at the denatured leather while believing that we are uncovering the vital secrets of the skin. Even this is something of an oversimplification because one is usually examining properties in a state somewhere between these two extremes. One can only find out where — and thus validate extrapolations to the situation *in vivo* — by carefully and systematically assessing the effect of any technique of preparation on the system being studied.

It is often argued that if one has a phenomenon called 'oxidative phosphorylation' and it occurs in a tissue homogenate – as in a mitochondrial preparation – one is entitled to study it in the latter. Here we must make distinctions between two different statements. Firstly, that particles one separates in the fraction can carry out oxidative phosphorylation as defined, and, secondly, that the particles contained in this fraction are those which have this biochemical activity *in vivo*. The difficulty is that one asserts the existence of the phenomenon by *quantitative* chemical assays of the different fractions which one compares. One gives lip-service to the knowledge that each of the fractions will be altered chemically to different extents by the preparation and then completely ignores this consideration in one's conclusion.

In general, it can be said that the less energy a separation or analytic technique involves, the less likely it is to alter the system being examined. Since, however, we are much more concerned with the dangers of temperature rise than heat generation, it would be appropriate to attempt to define four characteristics of biochemical technique which should be measured if possible and stated in publications. These are (i) the *rate* of temperature rise occurring during a procedure; (ii) the maximum temperature reached at each stage. These may be very difficult to measure at the sites at which they occur, so that one might have to use (iii) the *rate* of increase of heat generation, and (iv) the maximum rate of heat generation, at each stage.

Even without such measurements, we may rank the techniques in view of our separate examination of each, in increasing order of their probable generation of heat and consequent undesirability: radioactive techniques < chromatography < electrophoresis < histochemistry < subcellular fractionation < electron microscopy. This exercise is often not possible as the techniques are not necessarily alternatives.

What are the consequences of failing to carry out relevant control experiments? Why should one not believe that one knows what happens in mitochondria when one separates them by a technique which permits their identification in a fraction? Why then should one not continue to marvel at the 'endoplasmic reticulum'? Why not take the 'rough' with the 'smooth'?

I believe there are two cogent reasons. The first one is uncertainty. In this sense, I am using the word for two purposes, firstly because we like to be able to believe within ourselves what we proclaim to students. We must ask whether our statements imply several assumptions, which are contrary to the Law of Conservation of Energy, and several others, which are contrary to reason.

The second reason is as follows. If the effects of the separation techniques should prove to be large, it may turn out that one can *never* find out about the locations of enzymes in cell organelles *in vivo*, for example, or the relation between the lipid and the protein in the cell membranes. That is to say, if the amount or rate of injection of energy necessary to separate the mitochondria from the rest of the cell is such that the energy itself would alter the chemical nature of these constituents substantially, then it may be impossible by such energetic techniques *ever to know* the location of the enzymes *in vivo*. Clearly, if necessary measurements show that these secrets cannot be known, we are duty bound to abandon the techniques immediately. They would be unfruitful and a terrible waste of resources. In this circumstance no amount of repetition of the techniques or parrying of awkward questions at scientific meetings will ever validate them. It is the apprehension that biochemists may be travelling in the realm of uncertainty that has induced this book.

I would like to stress emphatically that it may well be that what is being taught to students today is largely correct. But I hasten to insist even more vehemently that until biochemists are prepared to control their experiments with the usual simple procedures which they preach, they may be misleading science as well as their students.

On a more practical note, it seems legitimate to wonder whether the singular lack of success that biochemists have enjoyed in characterizing natural carcinogens, neurotrophic factors or the active principles of embryo extracts, etc., may not be explained most easily by their employing techniques which degrade the active principles very early on in the separations. It seems to be no coincidence that the separations of 'nerve growth factors', serum globulins, embryo extracts etc. – which possess strong but labile biological activities – are usually done by evaporations, salt fractionations, chromatography, etc. — all of which are the relatively less energetic techniques.

One is continuously stressing to students the importance of control experiments yet, as far as the author is aware, no systematic attempt has ever been made to control any of the everyday biochemical procedures such as homogenization, centrifugation, subcellular fractionation, histochemistry, electronmicroscopy, etc. Conclusions arrived at in the absence of the assessment of the techniques should be regarded as 'not proven'.

A reasonable philosophy of science would be that the onus of unequivocal proof of a scientific assertion should rest firmly on the shoulders of those who assert it. Previous failure to do such control experiments, resulting in the pressure of the vast weight of findings without controls seems, in this age, no justification whatsoever for continuing to accept the validity of the experiments.

10

Practical conclusions

We may put forward a few practical suggestions for the jobbing research biochemist making quantitative measurements intended to be relevant to a situation *in vivo* (see definition of biochemistry, (d), page 103).

1. Techniques involving homogenization, centrifugation, and sonication – all of which generate heat – should be avoided if possible. If unavoidable, they should be done in pre-cooled conditions, minimally, and as slowly as possible.
2. Open systems, with small quantities of materials in thermally well-conducting vessels, are to be preferred to isolated or closed systems, especially with bulk materials.
3. Quantitative experiments should never be done without adequate recovery measurements and calibration curves in the presence of tissue.
4. 'Physiological' concentrations, pHs, co-factors, temperatures, and other conditions, should be simulated wherever possible.
5. If the effect of a technique is to change the system randomly more than the difference being detected, that technique produces uncertain results and should be abandoned.
6. Emotive words, like 'contamination', 'optimum', 'specific', 'ageing', 'mild' should be avoided in descriptions of experiments. The actual measurement or operation should be stated (see page 115).
7. Assumptions inherent in any technique whose use is contemplated should be systematically listed. If they have not been tested satisfactorily by earlier workers, the novice should examine them *before* embarking on the technique. If assumptions have not been examined at all, or cannot be for

practical or theoretical reasons, they should nevertheless be recorded in publications, even if distinguished pioneers have omitted to do so.
8. Verbal evasions of questions, citations of similarly behaving authority, and expressions like 'I would think such effects are minimal' are no substitute for properly controlled experiments.

11

Proposed strategy for biochemistry

The burden of this study is not that all experiments *in vitro* are suspect; rather that an experiment is only as good as its controls. If no controls have been done, which could have been done, the experiment is probably valueless. Biochemistry, which originally studied living tissues, has been carried away by an enthusiasm for physical techniques which *may* change the nature of the study so that what we discover is more a function of the method we use than the properties we seek to elucidate. In the author's opinion this has largely been the result of the application of high energy to unstable materials. We have learnt techniques from industrial chemists who use them for separating cream, examining ore crystals, or dyeing wool.

Our first task is to find out how far away from the living system we are, and I have suggested that an early step should be to study the biochemical changes that occur during dying and during the excision of tissues. This is fraught with difficulties because of the very rapid agonal changes; but until we can extrapolate back across this chasm of ignorance we will continue to study tissues that are running down – without knowing how far they have run.

Our second series of measures must be serious investigation of the effects of our techniques on the systems we are studying. This is a plea for scientists to use the elementary control experiments which are normally commended to schoolboys. It should not be necessary to make this point so long after Bacon, Harvey, Descartes and Fisher. We are attempting to measure the certainty of our methods upon which the validity of our findings depends otherwise unsupported. Furthermore, if we find that there is too much uncertainty due to a particular technique, we must take our courage in our hands and abandon that technique.

We can assess the uncertainty in a quantitative way. Though it will be tedious we should attempt to measure the rate of heat generation or temperature rise for each technique. We will know beforehand whether it can be used or not by comparing it with the known heat stability of biological materials. It is worth asking the question *en passant*: if a whole mammal cannot tolerate a body temperature of 42°C for long, why should its tissues?

Another aspect of the uncertainty is the degree to which the changes which occur during tissue preparation are reversible. These could be expressed as the ratio of the final to the original activity of the system. In this respect we may ask whether large molecules, other than proteins, 'denature'. The several physical measurements which shelter under this title could well be applied to an examination of whether similar partially reversible changes in the conformation can occur in carbohydrates, lipids, steroids, etc. It seems to be a useful working hypothesis to regard denaturation as a float in the cavalcade of dying. If we could reverse denaturation of tissues completely, we might jam the trigger of death, though this would probably have more unpleasant and unexpected effects than mankind has ever previously produced.

One way of avoiding uncertainty is to make intensive studies of natural completely unprocessed materials taken from the body before measuring their properties. Such examination could involve urine, saliva, cerebro-spinal fluid, and gastric juice, as well as 'blood, sweat and tears', which all contain extracellular fluids to which we have easy access. Instead of extracting enzymes from them and measuring their maximum activity at an extreme pH, we should be studying properties of naturally occurring enzymes in native mixtures at body pH and temperature. Furthermore in anaesthetized animals, we could study the ocular fluids, the synovial fluids, cartilage, bone, the uterus, the intestine, the kidney, etc., *in situ*.

In vitro there are, of course, many useful experiments that may be done. If a cerebral slice concentrates potassium ions ten times against the medium and requires energy to do so, or if an isolated ganglion transmits impulses, or if a separated perfused heart goes on beating, we are certainly wise to use them to study the mechanisms underlying these phenomena. We cannot, however, rest with the result unless we have explored and demonstrated its relationship to the tissue *in vivo*. With these preparations elicitation of as many of the 'functional' properties as we can that are found *in vivo* will immensely increase their value.

We may indicate other areas of experiment whose more active pursuit might appear advantageous. Firstly, we have the study of single cells, nuclei and giant axons, isolated by hand dissection, especially if they can be shown to retain excitability. Such techniques have been pioneered by Purkinje, Hyden, Pope and Lowry among others. Secondly, we have the artificial membranes made of natural and synthetic extracts by such people as Rudin and Bangham. Thirdly, we can define the properties of well characterized membranes so that we may have a basic physico-chemical knowledge about the behaviour of relatively simple systems; by comparing their properties with those of biological membranes *in vitro* and *in vivo* we can advance our understanding of the latter situation.

A further attitude to experiment which seems valuable is the scrutiny of the non-enzymic properties of biological systems. This is not quite the same as the controls for enzyme experiments. An enzyme is said to accelerate a reaction which would go very slowly indeed without it. It is therefore desirable to collect information about what dead tissue does to substrates of all kinds. This would give us the basic behaviour, deviation from which we can regard as the cunning of biochemistry.

There are two new systems of study which I would like to propose. One of them is the comparison of the behaviour of pure single amino-acids, proteins, carbohydrates, lipids, etc., in three distinct media. These three media are (i) as close a simulation as possible of the extracellular fluid, *e.g.* cerebrospinal fluid, plasma or Krebs-Ringer solution; (ii) a medium made up with the same constituents as the closest approximation we can design to the intracellular fluid; and (iii) a medium simulating the nucleoplasm as closely as possible. Comparison of the spectra, conductivity and charges of, say, albumen in these three salines at physiological pH, temperatures, and concentrations would throw light on how it might behave in the only conditions to which it is normally subjected. One would then gradually add more complex and esoteric constituents and build up a model of the physico-chemical behaviour of large biological molecules.

The other system I would like to mention is also new. If present beliefs about subcellular organelles prove misleading we can look for an alternative in Nature. Since we can see the nucleus, the cell membrane and the mitochondria under a light microscope, they cannot be soluble in the cytoplasm. I would, therefore, propose a simple separation system based on this fact. Briefly, one would disrupt the tissue by hand homogenization slowly in a large known volume of distilled water which would also act osmotically; this would depend on the absence of significant denaturation. The resultant mixture would then be filtered, if possible. The residue would be slowly evaporated to dryness at a temperature not exceeding 37°C to be reconstituted when needed.

This may be designated the insoluble fraction. The filtrate would be concentrated, also at a temperature not exceeding 37°C, to approximately its original volume and kept in a refrigerator. Slow filtration to separate the nuclei and, or mitochondria might be feasible, but the energy necessary to do so would have to be looked at very closely.

12
Summary and conclusions

A great deal of the modern biochemistry of tissues *in vitro* is done with unknowing disregard of the laws of thermodynamics and physics. I would conclude that the validation of some of the most popular techniques in world-wide use is grossly overdue. Until and unless this is done, all the findings based on them must be regarded as unproven. If the necessary control experiments should fail to validate the techniques, the techniques should be abandoned. At the moment biochemistry is in a state of uncertainty because elementary control experiments for complex procedures have *never* been done. I submit that they should and must be done soon.

Appendix 1

Misleading synonyms

	Term	Popular mis-use
(a)	Activation	Increased enzyme activity detected
(b)	Active transport	Concentration different than the medium
(c)	Ageing mitochondria	Properties changing with time
(d)	Biosynthesis	Uptake of radioactive precursor
(e)	Bound substances	Increased extraction by another agent
(f)	Contamination	Unwanted enzyme activity
(g)	Cytosol	Supernatant
(h)	Denaturation	Change of physical properties of protein
(i)	Enzyme activity	Breakdown of substrate
(j)	Extracellular space	Inulin, sucrose, or raffinose space
(k)	Fixation	Arrest of histological change
(l)	Inhibition	Decreased enzyme activity detected
(m)	Leached	Diffused
(n)	Membranes	Subcellular fraction
(o)	Mild centrifugation	1000 to 10,000 g
(p)	Non-specific uptake	Tissue uptake
(q)	Optimum enzyme activity	Maximum enzyme activity
(r)	Purification	Increased enzyme activity detected/mg protein
(s)	Recovery of activity	Percentage of total activity at end of preparation
(t)	Solubilization	Addition of detergents
(u)	Transport	Diffusion

The loose use of operational terms often obscures biochemical inexactitudes and assumptions. A few words must be said about some of them.

(a) **Activation** is often wrongly used to mean that the addition of a particular ion or a number of co-factors to an enzyme preparation increases the quantity of product detected. The apparently increased yield may arise from decreased stability of the substrate, acceleration of the enzyme reaction, increased substrate extraction, or alteration of optical properties of the whole system. The term should really be reserved for substances which have a stoichiometric relationship to the enzyme reaction.

(b) **Active transport** is not merely the concentration or exclusion of a substance by the tissue. It must have been shown to be against the electrochemical gradient and dependent on the supply of energy. It is not just a partition of a substance in favour of one or other phase of a system.

(c) **Ageing** of 'mitochondrial' preparations is the change of their properties with time following their isolation. Its existence implies that the effects of the separation procedure outlast the separation itself; the preparations have not reached equilibrium during the earlier measurements following fractionation. Ageing could be a phenomenon similar to dying and its characterization would be useful if one believes in the biochemical significance of 'mitochondrial' preparations.

(d) **Biosynthesis** is often used as a synonym for the uptake of a known radioactive precursor. However, it should imply much more. The term only has meaning if it has been shown that there is significantly more uptake in the presence of substrate and oxygen than in its absence, that there is significantly more of the precursor in that pathway than 'non-specifically', and that the extractant takes out very little of any product not involved in the pathway postulated. In practice all of these conditions are difficult to satisfy in experiments on living animals.

(e) **Bound substances** like calcium ions, ribonucleic acid, or proteins, are often so characterized because strong acids, bile salts, detergents, or sonication yield larger quantities from a preparation. This ignores the real possibility that these agents themselves have a special affinity for the 'bound' materials or actually increase the enzyme activity themselves.

(f) **Contamination** is an emotive term implying that an enzyme activity which is believed to reside exclusively in a fraction in which one is not interested, is found in significant concentration in the fraction to which one has not directed one's attention. The belief behind the concept is that there are enzymes which occur exclusively in one organelle. Since the 'purity' of the fraction is measured by the absence of other enzyme activity, there is no certain way of knowing by these techniques whether there are specific marker-enzymes of particular organelles. Indeed, belief in them makes cellular biochemistry much neater but in view of all the assumptions upon which the belief is based (page 33), it is more in the realm of an aesthetic than an axiom. Here we have uncertainty.

(g) **Cytosol** is sometimes used as a synonym for the supernatant and such mis-use implies that any soluble enzyme present in *any* cellular organelle *in vivo* would not diffuse into the supernatant during the preparation (see page 34, (g)).

(h) **Denaturation** can be used to mean changes in optical, physical or chemical properties (Joly, 1965) – if we may risk such dangerous distinctions. Clearly, in a description of techniques one must specify which particular technique is being used since they each may give different answers.

(i) **Enzyme activity** is not the same as the ability of the tissue to cause breakdown of substrate. The latter is a function of the total chemical environment of the substrate, including the presence in the preparation of activators, inhibitors, co-enzymes, a source of energy, etc., as well as the pure enzymes (see page 30).

Many substrates, such as cytochrome, or succinate, are very unstable themselves. Lactate and ATP, for example, absorb light. Their instability is increased by warming to 37°C. Any oxygenated mixture containing ions, buffers, and tissue, is liable to break down some substrate. We must draw a clear distinction between two circumstances. Firstly, the activity on a substrate which a tissue *in vivo* has at a particular moment in the absence of chemical addition of historical accident; this is a single, not necessarily controlled observation and its results cannot be attributed only to enzyme activity; secondly, we have a preparation which we have controlled for all non-enzymic activity by doing relevant preliminary and parallel experiments; in this case, we must subtract the non-enzymic activity from the total activity as measured in the complete preparation.

(j) **Extracellular space**. Research workers have claimed since the 1940s to measure extracellular space with inulin, sucrose, thiocyanate, raffinose and chloride. There was originally no evidence that they remained extracellular although their tissue concentrations are less than the medium concentrations. It was once supposed, and subsequently believed, that this itself constituted evidence that they did not enter cells. They could be degraded by tissue, react with it, or be repelled by it. In deference to the latter considerations, the original use of the term 'extracellular space' when inulin or sucrose space had been measured has recently given way to the expressions 'inulin' or 'sucrose spaces'; but their proponents still imply that they measure extracellular space, otherwise they would have abandoned using them.

(k) **Fixation** implies that because one has arrested obvious microscopic change, all biochemical and physico-chemical change within the system has been virtually halted, except in certain histochemical experiments where the fixative obligingly spares the enzymes we are studying (page 41).

(l) **Inhibition** has a well understood meaning in classical enzymology which is much more than the decrease of product due to the addition of a small quantity of inhibitor (please see 'activator' above). A specific inhibitor is presumably one that in low concentration decreases the rate of *only one* reaction stoichiometrically. The idea is usually taken to imply that it would not inhibit any other pathway significantly. This belief is often based on an extrapolation from perhaps two or three other reactions of the many thousand of which any viable tissue is capable. Clearly, this is statistically unacceptable. Just because an inhibitor affects a particular reaction in a homogenate furnished to exhibit one enzyme activity at its maximum rate, it does not mean that it would not affect other reactions of the same homogenate incubated under other conditions; nor does it mean *a fortiori* that in a more organized tissue, such as an isolated organ or a whole animal, the inhibition would be at the same site, as specific, or not counteracted by regulatory mechanisms. This latter extrapolation cannot be permitted with the knowledge that aspirin 'uncouples' oxidative phosphorylation extremely effectively.

(n) **Membranes.** The use of the word 'membranes' in respect of a subcellular fraction in which membrane fragments can be identified histologically should mean that the largest proportion *by volume* of the material in the fraction originates from the cells' membranes; further, that there is no significant contribution from other material which cannot be identified on electron microscopy or which has similar enzymatic activity. Often, however, it means only that in the particular fraction isolated by a well-defined series of steps, a particular particle or organelle is most commonly identified, not that most of the mass of the fraction is that particular organelle. This may be alright with mitochondria and nuclei, but it is very doubtful with more amorphous fractions containing 'endoplasmic reticulum', or 'microsomes'.

In separations of 'neuron' and 'neuroglial' fractions this distinction between a fraction of a pure cell and the only one in which that kind of cell can be identified, is clarified by the use of the words 'neuroglial fraction' or 'neuroglial-rich' fraction. Whereas the former would be expected to have the biochemical characteristics of neuroglia alone, the latter would not. In respect of the more amorphous subcellular fractions, there is no reason whatsoever to believe that much of the material seen does not originate from the breakage of other organelles. Individual particles are too small and irregular to compare with structures seen in electronmicrographs.

(o) **Mild centrifugation**, meaning 1,000 to 10,000 g, is to be compared, presumably, with 100,000 g. Among other consequences of the former treatment would be a pressure of 1 to 10 atmospheres (see pages 19-29). The use of the adjective 'mild' implies that it would have little or no effect on the physicochemical properties of the tissue so treated.

(p) **Non-specific uptake** of radioactive precursors, for example, is often used as an explanation for an uptake in which the research worker is not interested. It may mean only that the substance being studied is partitioned between the medium and the tissue in favour of the latter in the absence of an energy source.

(r) **Purification** of an enzyme may mean either that as a result of a preparative procedure a higher concentration of enzyme activity can be measured than before, or that non-active material has been separated from the enzyme preparation. Another interpretation would be that some steps in the preparative procedure might have *increased* the activity themselves, for example, by relocating co-factors, by generating heat, by changing the protein conformation, etc. The opposite may be said of loss of enzyme activity during preparation (see page 39).

(s) **Recovery of enzyme activity** following subcellular fractionation usually means the total enzyme activity of the subcellular fractions as a proportion of the activity of the crude homogenate. When this proportion is low the enzyme activity of a particular fraction is often expressed as a percentage of the sum total of all the activities in the final fractions added together. This has been discussed (see page 39). What the recovery should aim to measure is the total of activity in all the final fractions as a proportion of the tissue *before it was homogenized*. A method for doing this has been suggested (see page 16).

(t) **Solubilization** of a fraction often means the addition of a synthetic or natural detergent. When tissue becomes soluble its chemical properties must have changed.

(u) **Transport** can be defined as movement of particles or molecules, but its use often implies a mechanism for the movement. Diffusion is one possible mechanism; there are many others (see Bayliss, 1959; Wilbrandt & Rosenberg, 1961; Harris, 1960). Diffusion does not require a carrier, a semi-permeable membrane, special affinity of a tissue constituent, or any other external machinery; therefore, Occams razor encourages us to adduce it as a mechanism for transport before we look for more complicated possibilities. By the same token, unless we have demonstrated unequivocably that it is *not* the means of transport, we should not attribute it to another.

Bayliss, L. E. (1959) *Principles of General Physiology*, Vol. 1, London, Longmans.
Harris, E. J. (1960) *Transport and Accumulation in Biological Tissues*, London, Butterworth.
Joly, M. (1965) *Physico Chemical Approaches to Denaturation of Protein*, New York, Academic Press.
Wilbrandt, W., & Rosenberg, T. (1961) *Pharmacol. Revs.*, 13, 109.

Appendix 2

Radial pressure distribution in a rotating liquid

Consider a liquid column rotating in a horizontal plane with constant angular velocity, ω

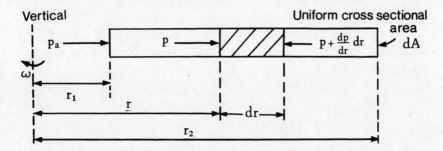

For the element shown shaded in the figure, Newton's Second Law reads:

$$pdA - \left(p + \frac{dp}{dr} \cdot dr\right) dA = \rho dA \cdot dr(-\omega^2 r)$$

which reduces to

$$p = \rho \frac{\omega^2 r^2}{2} + C_1$$

C_1 can be found by applying the boundary condition

$$r = r_1, \quad p = p_a$$

giving

$$p = p_a + \frac{\rho \omega^2}{2}(r^2 - r_1^2)$$

Notation for Appendices 2 and 3

- $2a$ diameter of particle
- r radius of particle path
- ω angular velocity of rotor
- s specific entropy
- C_p constant pressure specific heat
- T absolute temperature
- v specific volume $\equiv 1/\rho$
- ρ density $\equiv 1/v$
- p pressure
- β coefficient of volumetric expansion
- Δp a finite pressure change
- C_1, C_2 constants
- p_a atmospheric pressure

Reference for appendices 2 and 3 is Zemansky, M. Y. (1957) *Heat and Thermodynamics*, 4th edn., New York, McGraw Hill, page 246.

Appendix 3

Temperature change during a reversible adiabatic pressure change for a pure substance

Starting with the thermodynamic relationship

$$ds = C_p \frac{dT}{T} - \left(\frac{\partial v}{\partial T}\right)_p dp$$

we can write for an isentropic process (*i.e.* s = constant)

$$ds = 0 = C_p \frac{dT}{T} - \left(\frac{\partial v}{\partial T}\right)_p dp$$

Introducing β, the volumetric coefficient of expansion, defined by

$$\beta \equiv \frac{1}{v}\left(\frac{\partial v}{\partial T}\right)_p$$

we can write

$$C_p \frac{dT}{T} = v\beta\, dp$$

Now since β and C_p are only weak functions of pressure and only a *small* change in temperature is expected with a consequent *small* change in v, we can integrate the latter equation to give

$$\ln \frac{T_2}{T_1} = \frac{\beta v}{C_p} \Delta p$$

and

$$\frac{T_2}{T_1} = \exp\left\{\frac{\beta}{\rho C_p} \Delta p\right\}$$

Index

Absorption of light 19, 32, 33, 49, 59, 85
Acetone 44, 45
Acetylcholine 43, 48, 100
Acetyl thiocholine 48
Acetylcholinesterases 40
Acid phosphatase 18
Acid ribonuclease 18
Activation 115
Active transport 116
Acid alcohol 49
Adenosine deaminase 10
Adiabatic pressure, temperature change in 123
Adrenal cortex 1
Adrenalin 3, 4, 94
Aerodynamic heating 28
Affinity 32, 51, 58, 70, 76
Ageing of mitochondria 116
Alcohol 45, 47, 49, 50, 57, 63
Alcohol dehydrogenase 18
Aldehydes 54, 59
Alkali ribonuclease 18
Amidoschwarz 92
Amino-acids 4, 8, 10, 11, 14, 34, 39, 43, 56, 92, 112
AMPase 18
Anaesthetics 4, 5
Anions 10
Araldite 59
Arginase 10
Arginine phosphate 94
Artefacts 64, 98, 99
Artificial membranes 112
Aspartate amino-transferase 43
Aspirin 118
Assumptions inherent in chromatography 92; electronmicroscopy 63; electrophoresis 85; histochemistry 49; radioactive isotope techniques 73; subcellular fractionation 33, 38
ATP 5, 94, 117
ATPase 4, 39
Autoanalysis 92
Autolysis 42
Autoradiography 72, 75
Axons, giant 4, 112

Bacon, Lord 110
Bacteria 39
Bangham, A.D. 112
Bile salts 14, 33, 35
Binding 51, 116
Biosynthesis 14, 69, 116
Blood 68
Blood flow 68, 74
Bone 111
Brain 4, 55, 68
Buffer 14, 49, 81, 89, 117
Butanol 45

Calcium ions 34, 35, 62
Calibration 33, 64, 72, 77, 84
Carbohydrates 63, 91, 111, 112
Carbon dioxide, partial pressure 3
Carboxylic acids 56, 59
Carcinogens 106
Carcinoma 50
Carriers for chromatography 90
Cartilage 111
Catalase 10, 57
Catecholamines 3, 4, 18
Centrifugation 11, 18, 19-30, 92, 96, 119
Chloride 62, 117
Chlorophyll 62
Cholinesterase 48
Chromatography 86, 89-95, 101, 105
Clearing agents 49
Clinical measurements 103
Clove-oil 49
Cobalt 35
Cofactors 14, 70, 87, 99, 108
Column chromatography 91
Contamination 108, 114, 116
Control experiments 39, 52, 64, 70, 73, 76, 78, 101, 105
Cooling 16
Counting 72-3
Cryostat 44
Cytochrome 117
Cytochrome oxidase 44, 46, 64
Cytoplasm 8, 10, 13, 14, 20, 36, 43, 55, 56, 57, 59, 61
Cytosol 117

Davenport 48
Definitions of biochemistry 103-7
Dehydration 45-6, 54, 55, 56, 58, 91, 99
Denaturation 6, 25, 26, 32, 46, 50, 84, 86, 87, 111, 112, 117
Deparaffinization 47
Descartes 110
Detergents 14, 35
Diastase 23
Diffusion 42, 44, 48, 50, 69, 83, 120
Dilution 8, 10, 96, 97
Diptheria toxoid and anti-toxin 86
DNA 14, 19, 32, 58
Dying 2, 3, 4, 50, 96, 111

EDTA 14, 33, 35
Electron diffraction 59
Electronmicroscopy 38, 54-66, 86, 92, 99, 100, 105, 118, 119
Electrophoresis 79-88, 91, 92, 96, 101, 105
Elution 84, 85, 91
Embedding 45
Embryo extracts 106
Endoplasmic reticulum 55, 57, 59-61, 98, 105, 119

Index

Enzymes Preface, 6, 14, 15, 16, 17, 19, 20, 25, 30, 35, 36, 37, 43, 47, 50, 51, 83, 84, 91, 97, 98, 99, 106, 111, 112, 117, 118
Erythrocytes 20
Esterases 10, 44, 46
Esters 59
Ethers 47, 59
Ethylene glycol 45
Evaporation 97
Excretion 69
Exercise 2, 3, 5
Exoenzymes 39
Extracellular space and fluid 11, 13, 14, 56, 60, 71, 112, 117, 118
Extraction 31-2, 79-80
Extranuclear space 60, 61

Faraday's Law 82
Fats 91
Fatty acids 56
Fibroblasts 67
Fisher, Sir Ronald 110
Fixation 42, 44, 54, 96, 99, 118
Formalin 44, 50
Fractions 119
Freeze-drying 45-6
Freezing, rapid 6, 42, 50
Freeze substitution 45-6, 55
Friction 12, 14, 20

Ganglion cells 43, 62, 111
Glucose 4, 10, 49, 75
α-Glucosidase 17
Glucose-6-phosphatase 18
Glutamate 9
Glutamate dehydrogenase 57
Glutaraldehyde 63
Gold particles 62
Golgi body 60, 100

Hair 103
Halides 59
Harvey, William 110
Heart 111
Heat: conductivity 6, 13, 15, 19, 81, 97; dissipation or generation 12, 14, 15, 16, 18, 20, 28, 29, 80, 81, 83, 84, 93, 96, 97, 101, 108; radiation 59; solution 42; solvation 79; specific 13, 17, 24
HeLa cells 67
Histochemistry 41-53, 60, 99, 100, 105
Homogenate 13, 98, 119
Homogenization 8, 9, 11, 12-17, 96, 97, 112
Hormones 4, 56, 81
Hyden, H. 112
Hydrocarbons 59
Hydrogen ion 62, 77
Hypothermia 6
Hypoxia 5, 9

Incorporation 69, 71
Incubation 47

Indium trichloride 58
Inhibitors 47, 48, 70, 96, 118
Intestine 111
Intracellular fluid 112
Inulin 117
Invertase 17
Iodine 67
Iron 35
Isoenzymes 87
Isotopes, radioactive 67, 68, 69, 70, 74

Kidney 54, 55, 68, 111
Killing 1-5, 97

Lactate 3, 117
Leather 64, 103, 104
Light microscopy 56, 57, 58, 65, 85, 92, 98, 99, 112
Lipids 10, 47, 55, 56, 58, 63, 79, 106, 111, 112
Liver 55, 68
Lowry 112
Lyophilization 10
Lysozomes 100

Magnesium ions 34, 35, 62
Manganese 35
Membranes 6, 8, 13, 20, 54, 55, 98, 112, 118
Methacrylate 59
Methanol 45
Microbodies 54
Microsomes 16, 119
'Mild' reagents 119
Mitochondrion 6, 12, 16, 18, 20, 32, 54, 55, 57, 59, 98, 106, 112, 113
Muscle 68

Napthyl acetate 48
Native enzyme activities 111
Nerve growth factor 81, 94
Neurons 44, 62, 119
Neurotrophic factors 106
Nickel 35
Ninhydrin 92
Nitella 67
Non-enzymic activity 38
Non-specific uptake 70, 76, 119
Noradrenalin 3, 94
Nuclear membrane 61
Nucleoplasm 10, 14, 112
Nucleoside phosphorylase 10
Nucleotides 79, 86, 91
Nucleus 12, 13, 16, 20, 43, 55, 59, 60, 61, 62, 98, 112, 113

Ocular fluids 111
Oocytes 61, 62
Optical methods 18, 19, 33
Organic polymers 59
Organic solvents 10
Organo-metal complexes 56
Osmic acid 54, 55, 58, 63, 99, 100

Oxidative phosphorylation 4, 10, 105, 118
Oxygen, partial pressure 3, 4; in tissue 4, 49, 70; uptake 11

Pancreatic cells 62
Paper chromatography 91
Partition coefficient 30
Pepsin 23
Permanganate 54
Petrol ether 47
pH 3, 24, 32, 47, 54, 69, 80, 89, 90, 108, 111, 112
Phosphatases 41, 44, 46
Phosphate 68, 75, 77
Phosphocreatine 4, 5, 11, 86
Phosphotungstic acid 58, 63
Poliomyelitusis virus 86
Pollen 64
Polyethylene glycol 45
Polystyrene 59
Pools 69
Pope, A. 112
Pores 61, 98
Porphyrins 62
Potassium ion 9, 11, 14, 34, 42, 43, 62, 77, 104, 111
Precursor 71, 73, 74, 75
Pressure during centrifugation 20, 22, 23
Pressure in liquid 22, 25, 26, 121
Propylene glycol 63
Proteins 10, 14, 15, 25, 40, 55, 56, 62, 63, 79, 83, 86, 87, 90, 91, 97, 103, 106, 111, 112
Purity 73, 81, 119
Purkinje, J. E. 112
Pyruvate decarboxylase 6, 18

Quenching 72

Radial pressure distribution in rotating liquid 121
Radiation effects 58, 59, 68
Radioactivity measurements 67-78, 100
Raffinose 117
Recoveries 39, 73, 86, 92, 119
Respiration, artificial 5
Respiration of tissue 4
Retina 4
Ribonucleoproteins 57
Ribosomes 98
Rotor 22, 27
Rudin 112
Rumford, B. Preface, 14

Salivary gland cells 61
Sarcoplasmic reticulum 59, 99
Second Law of Thermodynamics 36, 51
Sedimentation coefficient 22
Separation 82, 105
Shrinkage of tissue 45, 55, 56, 99
Skeleton 68
Skin 103, 104

Sodium ion 4, 10, 14, 34, 42, 43, 62, 73, 77, 104
Solubilization 17, 120
Solvents 90-1, 96, *see also* organic solvents
'Somes' 100
Sonification 17
Specific activity 69
Specific resistance 80, 83
Sperm 6, 91
Spleen 68
Spores 91
Steroids 4, 103, 111
Strain energy 27
Streaming 60
Stress 1, 2, 5, 74, 96
Subcellular fractionation 1-40, 86, 98, 100, 105
Substrate 30, 47, 50, 99
Succinate 117
Succinate reductase 18
Succinic dehydrogenase 44, 46, 64
Sucrose 8, 9, 10, 11, 14, 15, 17-19, 26, 30, 117
Sulphatase 44
Supernatant 10
Swelling 8, 9, 10
Synapses 62
Synaptic vesicles 100
Synovial fluids 111

Temperature: coefficient 6, 18; rise 15, 16, 18, 24, 28, 58, 87, 89, 93, 97, 105, 108, 123
Tetanus toxin 86
Thiocyanate 9, 117
Toxins 91
Toxoid 86
Tissue cultures 59, 67
Transport 6, 120
Tritium 77
Trypsin 23
Turnover time 69

Unsaturated fatty acids 55
Uranyl acetate 58
Uterus 111

Vacuum 58
Vesicular bodies 100
Viscosity 14, 16, 18, 19, 20, 22, 26
Vitamins 56

Water 8, 15, 16, 17, 55, 63, 68
Wood 64, 103
Wool 64, 103

X-ray diffraction 58
Xylem 103
Xylol 47, 49

Zinc 35

666-2334 Library
1648 Books